Interactive Student Edition

Reveal MATH™

Course 2 • Volume 1

McGraw Hill Education

Cover: (l to r, t to b) McGraw-Hill Education, luismmolina/E+/Getty Images, Blend Images - Erik Isakson/ Brand X Pictures/Getty Images, Atomic Imagery/DigitalVision/Getty Images

my.mheducation.com

Send all inquiries to:
McGraw-Hill Education
STEM Learning Solutions Center
8787 Orion Place
Columbus, OH 43240

ISBN: 978-0-07-667374-2
MHID: 0-07-667374-X

Reveal Math, Course 2
Interactive Student Edition, Volume 1

Printed in the United States of America.

2 3 4 5 6 7 8 9 10 11 LMN 28 27 26 25 24 23 22 21 20 19

Contents in Brief

Reveal Math™ Makes Math Meaningful...

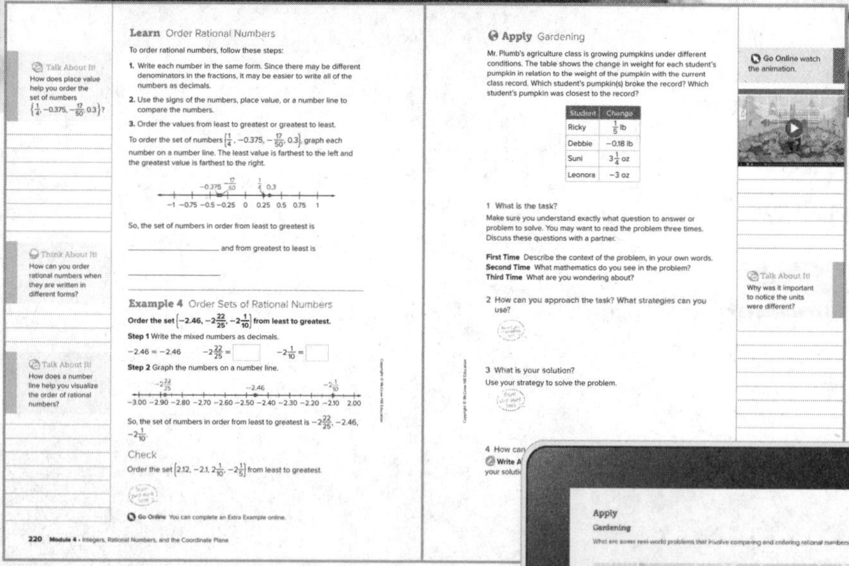

Learn Order Rational Numbers

To order rational numbers, follow these steps:

1. Write each number in the same form. Since there may be different denominators in the fractions, it may be easier to write all of the numbers as decimals.
2. Use the signs of the numbers, place value, or a number line to compare the numbers.
3. Order the values from least to greatest or greatest to least.

To order the set of numbers $\left\{\frac{1}{4}, -0.375, -\frac{17}{50}, 0.3\right\}$, graph each number on a number line. The least value is farthest to the left and the greatest value is farthest to the right.

So, the set of numbers in order from least to greatest is

________________ and from greatest to least is

____

Talk About It!
How does place value help you order the set of numbers $\left\{\frac{1}{4}, -0.375, -\frac{17}{50}, 0.3\right\}$?

Think About It!
How can you order rational numbers when they are written in different forms?

Example 4 Order Sets of Rational Numbers

Order the set $\left\{-2.46, -2\frac{22}{25}, -2\frac{1}{10}\right\}$ from least to greatest.

Step 1 Write the mixed numbers as decimals.

$-2.46 = -2.46$ $\quad -2\frac{22}{25} = \square$ $\quad -2\frac{1}{10} = \square$

Step 2 Graph the numbers on a number line.

So, the set of numbers in order from least to greatest is $-2\frac{22}{25}, -2.46, -2\frac{1}{10}$.

Check

Order the set $\left\{2.12, -2.1, 2\frac{1}{10}, -2\frac{1}{5}\right\}$ from least to greatest.

Talk About It!
How does a number line help you visualize the order of rational numbers?

Go Online You can complete an Extra Example online.

220 Module 4 • Integers, Rational Numbers, and the Coordinate Plane

Apply Gardening

Mr. Plumb's agriculture class is growing pumpkins under different conditions. The table shows the change in weight for each student's pumpkin in relation to the weight of the pumpkin with the current class record. Which student's pumpkin(s) broke the record? Which student's pumpkin was closest to the record?

Student	Change
Ricky	$\frac{1}{5}$ lb
Debbie	−0.18 lb
Suni	$3\frac{1}{4}$ oz
Leonora	−3 oz

1 What is the task?

Make sure you understand exactly what question to answer or problem to solve. You may want to read the problem three times. Discuss these questions with a partner.

First Time Describe the context of the problem, in your own words.
Second Time What mathematics do you see in the problem?
Third Time What are you wondering about?

2 How can you approach the task? What strategies can you use?

3 What is your solution?

Use your strategy to solve the problem.

4 How can

Write A

your soluti

Go Online watch the animation.

Talk About It!
Why was it important to notice the units were different?

Interactive Student Edition

Learning on the Go!

The flexible approach of *Reveal Math* can work for you using digital only or digital and your *Interactive Student Edition* together.

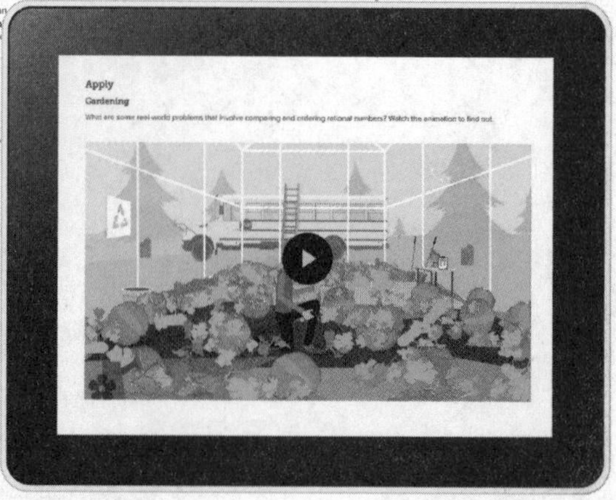

Student Digital Center

...to Reveal YOUR Full Potential!

Reveal Math™ Brings Math to Life in Every Lesson

Reveal Math is a blended print and digital program that supports access on the go. You'll find the *Interactive Student Edition* aligns to the Student Digital Center, so you can record your digital observations in class and reference your notes later, or access just the digital center, or a combination of both! The Student Digital Center provides access to the interactive lessons, interactive content, animations, videos and technology-enhanced practice questions.

Write down your username and password here

Username: ______________________________

Password: ______________________________

Go Online! my.mheducation.com

Web Sketchpad® Powered by The Geometer's Sketchpad®- Dynamic, exploratory, visual activities embedded at point of use within the lesson.

Animations and Videos – Learn by seeing mathematics in action.

Interactive Tools – Get involved in the content by dragging and dropping, selecting, and completing tables.

Personal Tutors – See and hear a teacher explain how to solve problems.

eTools – Math tools are available to help you solve problems and develop concepts.

TABLE OF CONTENTS

Module 1

Proportional Relationships

Essential Question

What does it mean for two quantities to be in a proportional relationship?

Module 2
Solve Percent Problems

Essential Question

How can percent describe the change of a quantity?

TABLE OF CONTENTS

Module 3

Operations with Integers

Essential Question

How are operations with integers related to operations with whole numbers?

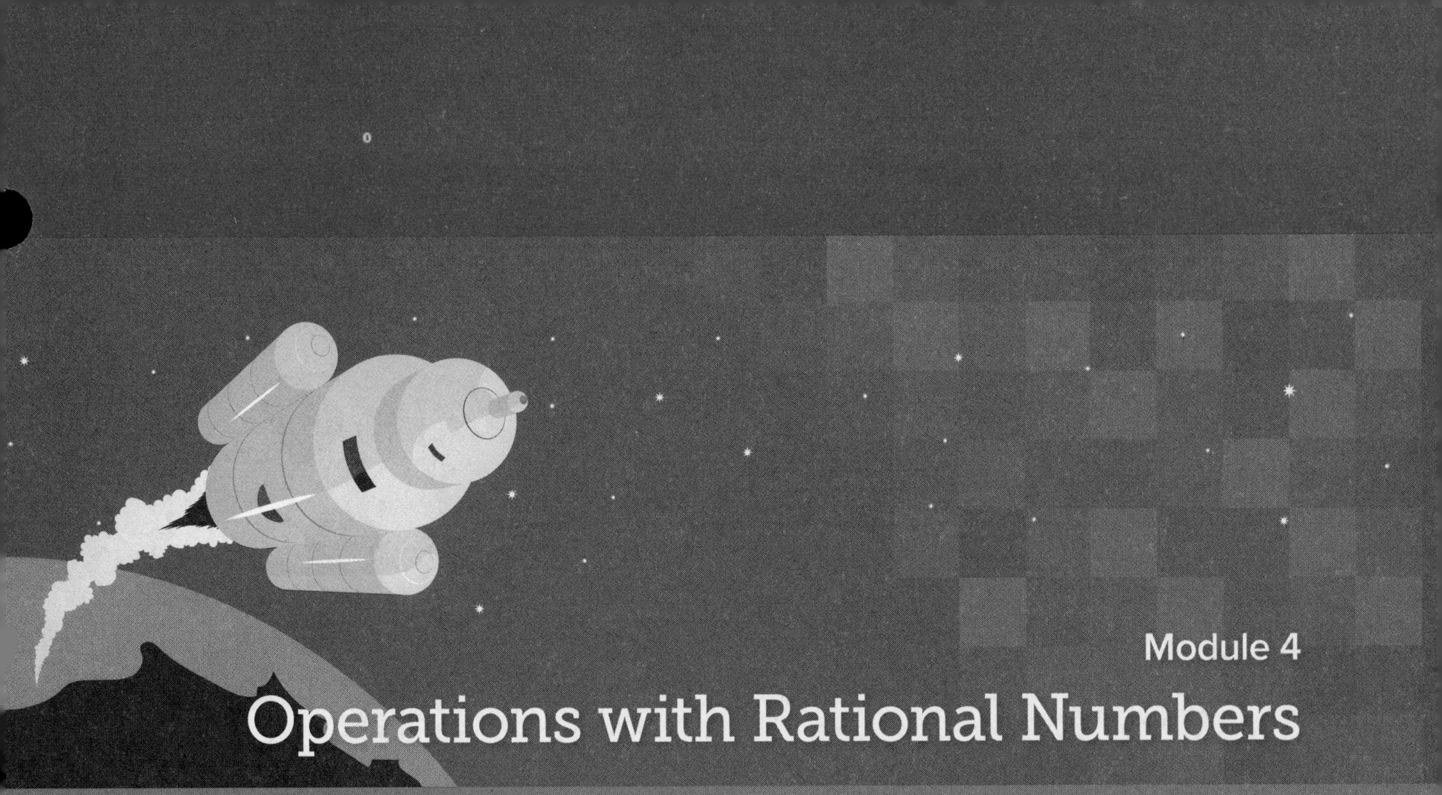

Module 4

Operations with Rational Numbers

Essential Question

How are operations with rational numbers related to operations with integers?

TABLE OF CONTENTS

Module 5

Simplify Algebraic Expressions

Essential Question

Why is it beneficial to rewrite expressions in different forms?

Module 6

Write and Solve Equations

Essential Question

How can equations be used to solve everyday problems?

TABLE OF CONTENTS

Module 7

Write and Solve Inequalities

Essential Question

How are solutions to inequalities different from solutions to equations?

Module 8

Geometric Figures

Essential Question

How does geometry help to describe objects?

TABLE OF CONTENTS

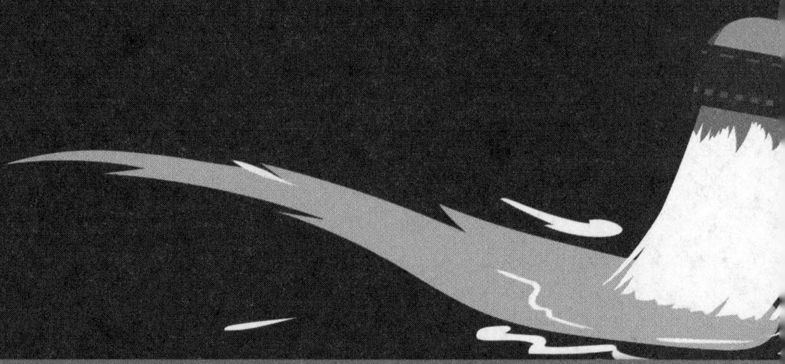

Module 9

Measure Figures

Essential Question

How can we measure objects to solve problems?

Module 10

Probability

Essential Question

How can probability be used to predict future events?

TABLE OF CONTENTS

Module 11

Sampling and Statistics

Essential Question

How can you use a sample to gain information about a population?

Module 1

Proportional Relationships

Essential Question

What does it mean for two quantities to be in a proportional relationship?

What Will You Learn?

Place a checkmark (✓) in each row that corresponds with how much you already know about each topic **before** starting this module.

KEY

■ — I don't know. ◆ — I've heard of it. ★ — I know it!

	Before			After		
	■	◆	★	■	◆	★
computing unit rates involving ratios of fractions						
determining whether a relationship is proportional by looking at a table of values						
determining whether a relationship is proportional by looking at a graph						
finding and interpreting the constant of proportionality						
interpreting the points (0, 0) and (1, r) on the graph of a proportional relationship						
representing proportional relationships with equations						
solving problems involving proportional relationships						

Foldables Cut out the Foldable and tape it to the Module Review at the end of the module. You can use the Foldable throughout the module as you learn about proportional relationships.

What Vocabulary Will You Learn?

Check the box next to each vocabulary term that you may already know.

☐ constant of proportionality ☐ proportional

☐ nonproportional ☐ proportional relationship

☐ proportion ☐ unit rate

Are You Ready?

Complete the Quick Review to see if you are ready to start this module.
Then complete the Quick Check.

Quick Review

Example 1
Write ratios.

Write the ratio of wins
to losses.

Mavericks	
Wins	10
Losses	12
Ties	8

wins : losses
 10 : 12

The ratio of wins to losses is 10 : 12.

Example 2
Determine if ratios are equivalent.

Determine whether the ratios 250 miles
in 4 hours and 500 miles in 8 hours are
equivalent.

$$250 \text{ miles : 4 hours} \rightarrow \frac{250}{4} \xLeftrightarrow{\div 2} = \frac{125}{2} \text{ or } 62\frac{1}{2}$$

$$500 \text{ miles : 8 hours} \rightarrow \frac{500}{8} \xLeftrightarrow{\div 4} = \frac{125}{2} \text{ or } 62\frac{1}{2}$$

The ratios are equivalent because the
ratio $\frac{125}{2}$ is maintained.

Quick Check

1. Refer to the table in Example 1. Write
the ratio of wins to total games.

2. Determine whether the ratios 20 nails for
every 5 shingles and 12 nails for every
3 shingles are equivalent.

How Did You Do?
Which exercises did you answer correctly in the Quick Check?
Shade those exercise numbers at the right.

① ②

Unit Rates Involving Ratios of Fractions

I Can... find unit rates when one or both quantities are fractions.

Explore Find Unit Rates with Fractions

Online Activity You will use bar diagrams to explore how to find a unit rate when one or both quantities of a given rate are fractions.

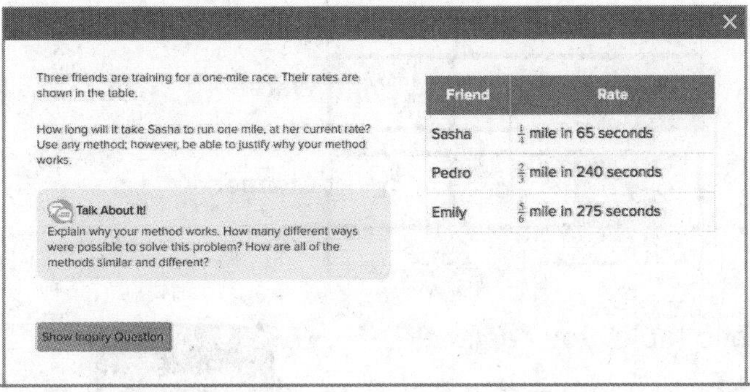

Three friends are training for a one-mile race. Their rates are shown in the table.
How long will it take Sasha to run one mile, at her current rate? Use any method; however, be able to justify why your method works.
Talk About It!
Explain why your method works. How many different ways were possible to solve this problem? How are all of the methods similar and different?
Show Inquiry Question
Friend
Rate
Sasha
$\frac{1}{4}$ mile in 65 seconds
Pedro
$\frac{2}{3}$ mile in 240 seconds
Emily
$\frac{5}{6}$ mile in 275 seconds

Learn Unit Rates Involving Ratios of Fractions

A recipe to make a diluted cleaning solution calls for 4 gallons of water mixed with $\frac{1}{3}$ cup of cleaner. The ratio of gallons of water to cups of cleaner is 4 to $\frac{1}{3}$ or $4 : \frac{1}{3}$. You have 1 cup of cleaner and you want to use all of it. How many gallons of water do you need to mix with 1 cup of cleaner to maintain the ratio $4 : \frac{1}{3}$?

To find the *unit rate*, the number of gallons of water needed to mix with 1 cup of cleaner, you can use various strategies.

What Vocabulary Will You Learn?
unit rate

Study Tip

You learned about ratios, rates, and unit rates in a previous grade.

A *ratio* is a comparison of two quantities, in which for every *a* units of one quantity, there are *b* units of another quantity.

A *rate* is a ratio that compares two quantities with unlike units.

A *unit rate* compares the first quantity per every 1 unit of the second quantity.

Your Notes

Talk About It!

Explain to a partner why the rate $4 : \frac{1}{3}$ is not a unit rate.

Study Tip

The expression $\frac{4}{\frac{1}{3}}$ is called a *complex fraction*. A complex fraction is a fraction in which the numerator or denominator, or both, are also fractions.

Talk About It!

How is finding a unit rate when one of the quantities is a fraction similar to finding a unit rate when both quantities are whole numbers? How is it different?

The double number line shows that, for every $\frac{1}{3}$ cup of cleaner, 4 gallons of water are needed. So, 12 gallons of water are needed to mix with 1 cup of cleaner.

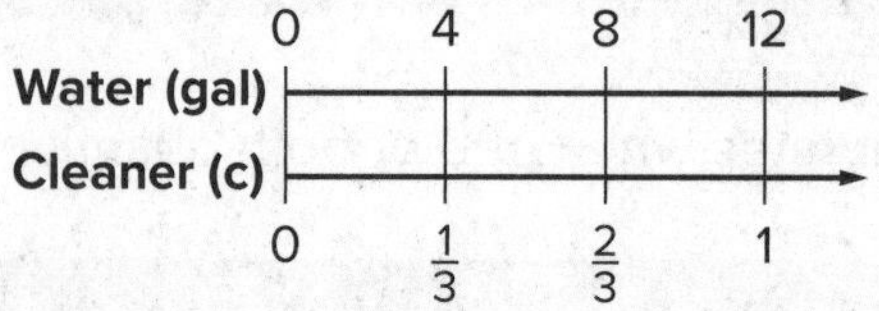

The bar diagram also shows that 12 gallons of water are needed to mix with 1 cup of cleaner. For every $\frac{1}{3}$ cup of cleaner, 4 gallons of water are needed.

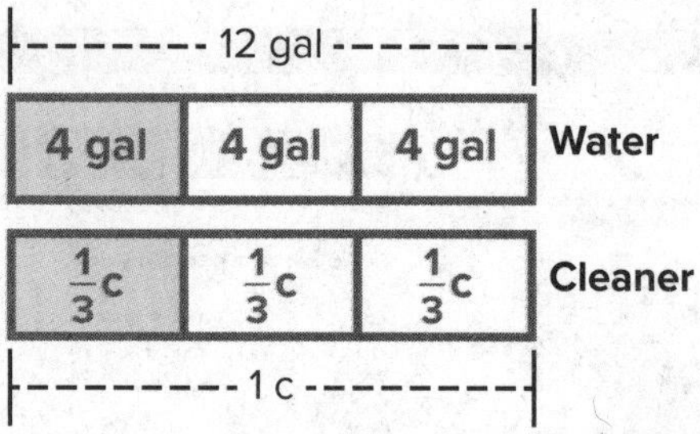

By creating a ratio table, you can scale forward by multiplying both $\frac{1}{3}$ and 4 by 3. The ratio table confirms that 12 gallons of water are needed to mix with 1 cup of cleaner.

	×3 →	
Cleaner (c)	$\frac{1}{3}$	1
Water (gal)	4	12
	×3 →	

You can also use division when finding a unit rate. Recall that a ratio or rate can be written in fraction form. The ratio 4 : 1 can be written in fraction form as $\frac{4}{1}$.

The ratio $4 : \frac{1}{3}$ can be written in fraction form as $\frac{4}{\frac{1}{3}}$.

Because a fraction bar indicates division, you can divide the numerator by the denominator to find the unit rate.

$\frac{4}{\frac{1}{3}} = 4 \div \frac{1}{3}$ — The fraction bar indicates division.

$= \frac{4}{1} \div \frac{1}{3}$ — Write 4 as $\frac{4}{1}$.

$= \frac{4}{1} \cdot \frac{3}{1}$ — Multiply by the multiplicative inverse of $\frac{1}{3}$.

$= \frac{12}{1}$, or 12 — Multiply the fractions.

So, $\frac{4}{\frac{1}{3}} = 12$.

Using any of these strategies, the unit rate is 12 gallons of water for every 1 cup of cleaner.

Example 1 Find Unit Rates

Tia is painting one side of her shed. She paints 36 square feet in 45 minutes.

At this rate, how many square feet can she paint each hour?

You know that 45 minutes is $\frac{3}{4}$ of an hour. So, Tia's rate is 36 square feet per $\frac{3}{4}$ hour. You need to find the unit rate, the number of square feet she can paint per 1 hour.

Method 1 Use a bar diagram.

Draw two bars to model the ratio $36 : \frac{3}{4}$. Divide each bar into four sections, because $\frac{3}{4}$ is a multiple of $\frac{1}{4}$, and there are 4 sections of $\frac{1}{4}$ hour in 1 hour.

To find the unit rate, first find the value of each section in the bar representing square feet. Because three sections have a value of 36 square feet, each section has a value of $36 \div 3$, or 12 square feet. Because $4(12) = 48$, the unit rate is 48 square feet per hour.

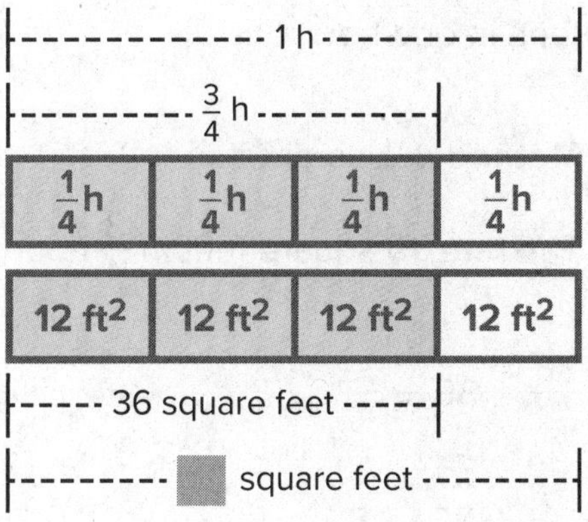

Method 2 Use a double number line.

The top number line represents the number of hours. The bottom number line represents the number of square feet. Mark and label the ratio $36 : \frac{3}{4}$.

Mark and label four equal increments of $\frac{1}{4}$ on the top number line. Mark the same number of equal increments on the bottom number line.

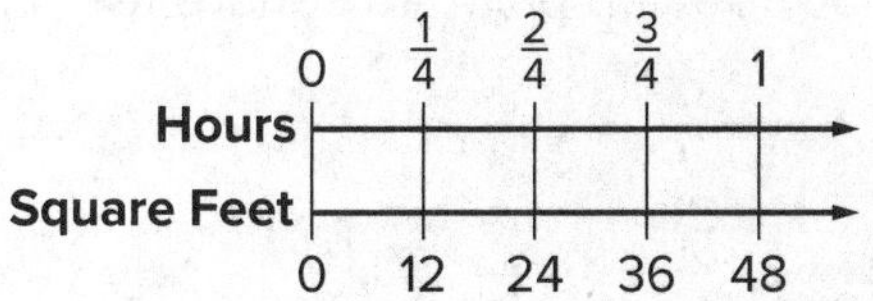

Each increment on the bottom number line represents 12 square feet. Because $4(12) = 48$, the unit rate is 48 square feet per hour.

(continued on next page)

Think About It!
Why is 45 minutes equal to $\frac{3}{4}$ of an hour?

Talk About It!
Use mathematical reasoning to explain why Tia can paint more than 36 square feet per hour.

Talk About It!
Why do you need to scale backward first before scaling forward?

Method 3 Use a ratio table.

The ratio table shows the number of square feet painted in $\frac{3}{4}$ hour. Scale backward to find the number of square feet painted in $\frac{1}{4}$ hour.

÷3

Hours	$\frac{1}{4}$	$\frac{3}{4}$
Square feet	?	36

÷3

Scale forward to find the number of square feet Tia can paint in 1 hour. This is the unit rate.

×4

Hours	$\frac{1}{4}$	$\frac{3}{4}$	1
Square feet	12	36	?

×4

Because $\frac{1}{4}(4) = 1$, multiply 12(4). So, Tia can paint 12(4), or 48 square feet, in one hour.

Talk About It!
Compare the four methods. What operation(s) did you use with each of the methods?

Method 4 Use division.

The rate 36 square feet in $\frac{3}{4}$ hour can be written as $\frac{36}{\frac{3}{4}}$.

$\frac{36}{\frac{3}{4}} = 36 \div \frac{3}{4}$	Write the complex fraction as a division problem.
$= \frac{36}{1} \div \frac{3}{4}$	Write 36 as $\frac{36}{1}$.
$= \frac{36}{1} \cdot \frac{4}{3}$	Multiply by the reciprocal of $\frac{3}{4}$, which is $\frac{4}{3}$.
$= \frac{144}{3}$ or 48	Multiply. The unit rate is 48 ft^2 per hour.

So, Tia can paint 48 square feet in one hour.

Check

Doug entered a canoe race. He paddled 5 miles in $\frac{2}{3}$ hour. What is his average speed in miles per hour? Use any strategy.

Show your work here

Go Online You can complete an Extra Example online.

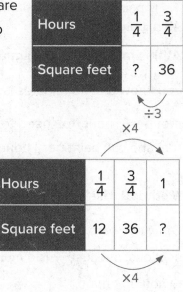

Example 2 Find Unit Rates

Josiah can jog $\frac{5}{6}$ mile in 15 minutes.

Find his average speed in miles per hour.

You know that 15 minutes is $\frac{1}{4}$ hour. So, Josiah's rate is $\frac{5}{6}$ mile per $\frac{1}{4}$ hour. You need to find the unit rate, the number of miles he can jog per 1 hour.

Method 1 Use a bar diagram.

Draw two bars to model the ratio $\frac{5}{6}:\frac{1}{4}$. Divide each bar into 4 sections, because there are 4 sections of $\frac{1}{4}$ in 1 hour.

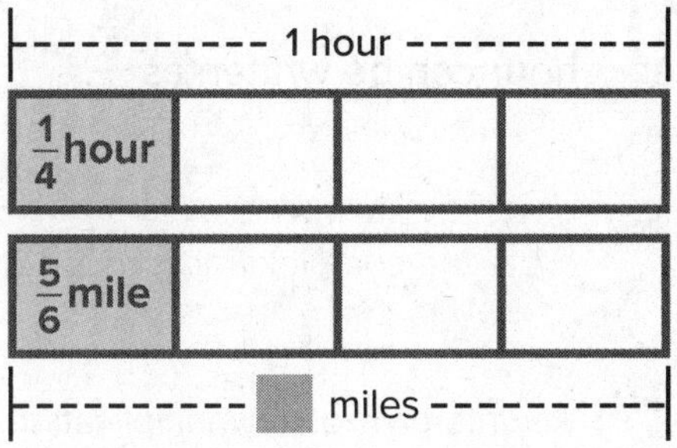

To find the unit rate, first find the value of each section in the bar representing miles. Each section has a value of $\frac{5}{6}$ mile. Because $4\left(\frac{5}{6}\right) = \frac{20}{6}$ or $3\frac{1}{3}$, the unit rate is $3\frac{1}{3}$ miles per hour.

Method 2 Use a double number line.

The top number line represents the number of hours. The bottom number line represents the number of miles. Mark and label the ratio $\frac{5}{6}:\frac{1}{4}$.

Mark and label four equal increments of $\frac{1}{4}$ on the top number line. Mark the same number of equal increments on the bottom number line.

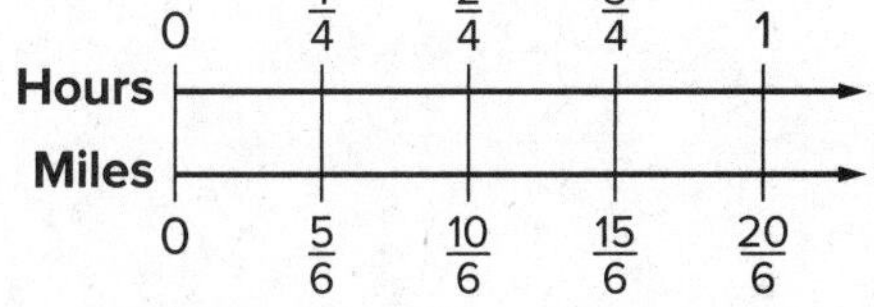

Each increment on the bottom number line represents $\frac{5}{6}$ mile.

Because $4\left(\frac{5}{6}\right) = \frac{20}{6}$, or $3\frac{1}{3}$, the unit rate is $3\frac{1}{3}$ miles per hour.

(continued on next page)

Think About It!

What do you notice about both quantities of the rate?

Method 3 Use a ratio table.

The ratio table shows the number of miles jogged in $\frac{1}{4}$ hour. Scale forward to find the number of miles Josiah can jog in 1 hour. This is the unit rate.

×4

Hours	$\frac{1}{4}$	1
Miles	$\frac{5}{6}$	?

×4

Because $\frac{1}{4}(4) = 1$, multiply $\frac{5}{6}(4)$. So, Josiah can jog $\frac{5}{6}(4)$, or $3\frac{1}{3}$ miles in one hour.

Talk About It!
How could you word this problem using minutes instead of hours? Does it change the outcome of Josiah's average speed?

Method 4 Use division.

The rate $\frac{5}{6}$ mile per $\frac{1}{4}$ hour can be written as $\frac{\frac{5}{6}}{\frac{1}{4}}$.

$\frac{\frac{5}{6}}{\frac{1}{4}} = \frac{5}{6} \div \frac{1}{4}$ Write the complex fraction as a division expression.

$= \frac{5}{6} \cdot \frac{4}{1}$ Rewrite division as multiplication.

$= \frac{20}{6}$ Multiply.

$= 3\frac{1}{3}$ Simplify. The unit rate is $3\frac{1}{3}$ miles per hour.

With each representation, because there are 4 quarter-hours in one hour, his average speed is found by multiplying by 4. So, Josiah can jog $3\frac{1}{3}$ miles per hour.

Check

A garden hose was left on in a yard and spilled $\frac{7}{8}$ gallon every $\frac{2}{3}$ minute. Find the average number of gallons spilled per minute. Use any strategy.

Show your work here

Go Online You can complete an Extra Example online.

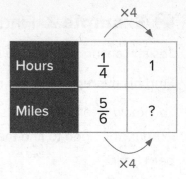

Apply Kayaking

Carolina and her friends are training separately for a kayaking competition. The average distance and time traveled by each is shown in the table. If the distance kayaked in the competition is 3 miles, predict who will win based on the rates shown. Predict how long it will take the winner to complete the race, if their rate remains constant.

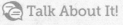

Person	Carolina	Leslie	Bryan	Javier
Average Distance (mi)	$\frac{7}{8}$	1	$\frac{3}{4}$	1
Average Time (h)	$\frac{1}{2}$	$\frac{1}{3}$	$\frac{3}{4}$	$\frac{2}{3}$

1 What is the task?

Make sure you understand exactly what question to answer or problem to solve. You may want to read the problem three times. Discuss these questions with a partner.

First Time Describe the context of the problem, in your own words.
Second Time What mathematics do you see in the problem?
Third Time What are you wondering about?

2 How can you approach the task? What strategies can you use?

Record your observations here

3 What is your solution?

Use your strategy to solve the problem.

Show your work here

4 How can you show your solution is reasonable?

Write About It! Write an argument that can be used to defend your solution.

Talk About It!
Which method is more advantageous to use when solving this problem?

Check

A walk-a-thon was held at a local middle school. The table gives the average distances and times for four walkers for certain periods of the walk-a-thon. If the total distance of the route was 2 miles, who completed the route first? How long did it take her to complete the route if she walked at a constant rate?

Person	Distance (mi)	Time (min)
Lakeisha	$\frac{7}{8}$	18
Baydan	$\frac{9}{10}$	22
Madison	$\frac{4}{5}$	17

Show your work here

Go Online You can complete an Extra Example online.

Pause and Reflect

Compare the process for finding unit rates involving fractions with what you know about dividing fractions. How are they similar? How are they different?

Record your observations here

Name ________________ Period ________ Date ________

Practice

Go Online You can complete your homework online.

Solve each problem. Use any strategy, such as a bar diagram, double number line, ratio table, or division.

1. A truck driver drove 48 miles in 45 minutes. At this rate, how many miles can the truck driver drive in one hour? **(Example 1)**

2. Russell runs $\frac{9}{10}$ mile in 5 minutes. At this rate, how many miles can he run in one minute? **(Example 1)**

3. A small airplane flew 104 miles in 50 minutes. At this rate, how many miles can it fly in one hour? *(50 minutes =* $\frac{5}{6}$ *hour)* **(Example 1)**

4. DeAndre downloaded 8 apps onto his tablet in 12 seconds. At this rate, how many apps could he download in one minute? *(12 seconds =* $\frac{1}{5}$ *minute)* **(Example 1)**

5. In Lixue's garden, the green pepper plants grew 5 inches in $\frac{3}{4}$ month. At this rate, how many feet can they grow in one month? *(Let 1 month = 4 weeks)* **(Example 2)**

6. Thunder from a bolt of lightning travels $\frac{1}{10}$ mile in $\frac{1}{2}$ second. At this rate, how many miles can it travel in one second? **(Example 2)**

7. The average sneeze can travel $\frac{3}{100}$ mile in 3 seconds. At this rate, how far can it travel in one minute? *(3 seconds =* $\frac{1}{20}$ *minute)* **(Example 2)**

Test Practice

8. Multiselect Anita is making headbands for her softball team. She needs a total of $\frac{3}{4}$ yard of fabric. Select all types of fabric that cost less than $8 per yard. **(Example 2)**

- ☐ cotton
- ☐ flannel
- ☐ fleece
- ☐ terry cloth

Fabric	Total Cost for $\frac{3}{4}$ Yard ($)
Cotton	5.54
Flannel	2.62
Fleece	4.27
Terry Cloth	6.52

Apply

9. During the first seconds after takeoff, a rocket traveled 208 kilometers in 50 minutes at a constant rate. Suppose a penny is dropped from a skyscraper and could travel 153 kilometers in $\frac{1}{2}$ hour at a constant rate. Which of these objects has a faster unit rate per hour? How much faster?

10. To prepare for a downhill skiing competition, Roman completed three training sessions. The table shows his average time and distance for each session. Did Roman's rate, in miles per hour, increase from session to session? Write an argument that can be used to defend your solution.

Session	Time (hr)	Distance (mi)
1	$\frac{2}{125}$	$\frac{9}{10}$
2	$\frac{3}{200}$	$\frac{11}{12}$
3	$\frac{3}{250}$	$\frac{17}{20}$

11. MP **Reason Abstractly** Explain why a student who runs $\frac{3}{4}$ mile in 6 minutes is faster than a student who runs $\frac{1}{2}$ mile in 5 minutes.

12. Compare and contrast the rates $\frac{4}{5}$ *mile in 8 minutes* and *4 minutes to travel* $\frac{2}{5}$ *mile.*

13. MP **Find the Error** Carli made 9 greeting cards in $\frac{3}{4}$ hour. She determined her unit rate to be $\frac{1}{12}$ card per hour. Find her error and correct it.

14. MP **Be Precise** A standard shower drain can drain water at the rate of 480 gallons in $\frac{2}{3}$ hour. Create three different rates, using the same or different units, that are all equivalent to this rate. Be precise in the units you choose. Then find the unit rate, in gallons per minute.

Understand Proportional Relationships

I Can... use models and ratio reasoning to understand how a proportional relationship can exist between quantities.

What Vocabulary Will You Learn?
proportional relationship

Learn Proportional Relationships

When baking, the ratio(s) of ingredients is important to maintain. The table shows a common recipe for pizza dough. Too much or too little of any one ingredient will not create a good dough.

Ingredient	Amount
Flour	3 c
Salt	$\frac{1}{2}$ tsp
Yeast	2 tsp
Water	1 c
Olive Oil	4 tsp

In this recipe, the ratio of cups of flour to cups of water is 3 : 1. Suppose you wanted to make two batches of dough. Is the ratio of flour to water the same? What if you wanted to make three batches?

The bar diagrams show the relationship between flour and water for different batches of dough. In each batch, 3 equal-size sections represent cups of flour, and 1 section of the same size represents cups of water.

One Batch

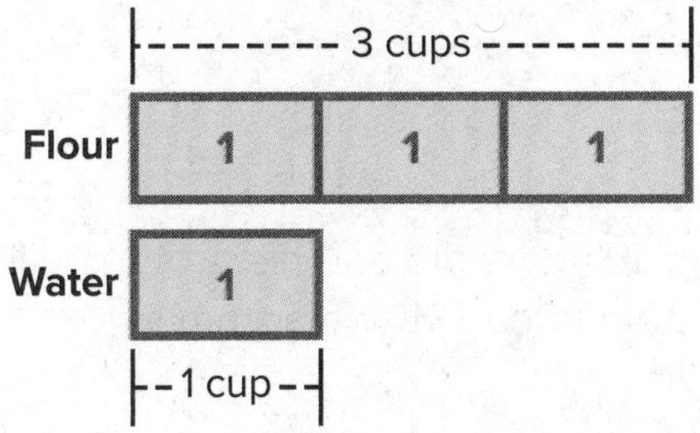

For one batch, the ratio is 3 cups of flour to 1 cup of water.

Two Batches

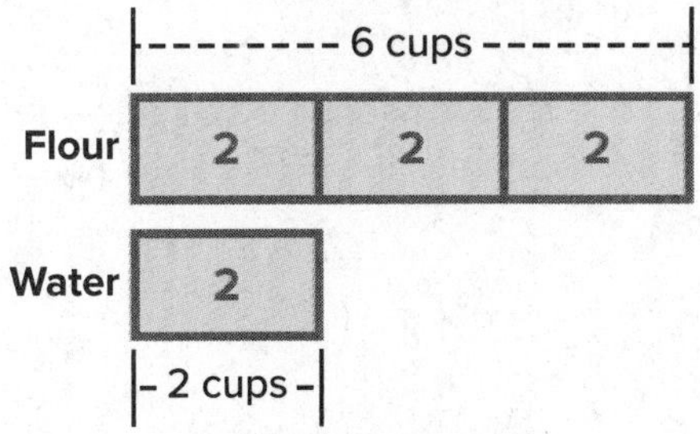

For two batches, the ratio is 6 cups of flour to 2 cups of water. For every 3 cups of flour, you need 1 cup of water. The ratio 3 : 1 is maintained.

Three Batches

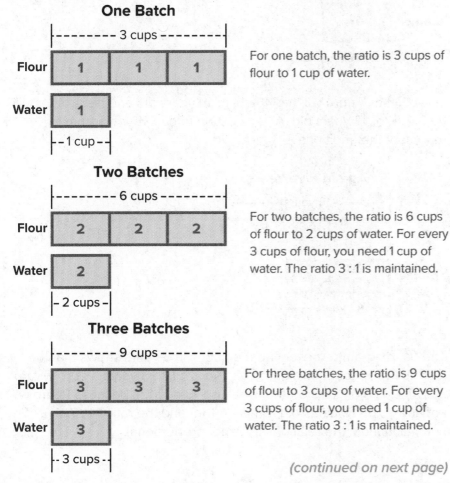

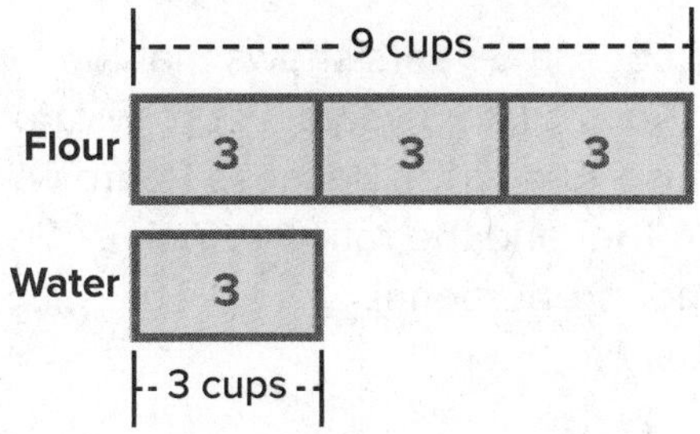

For three batches, the ratio is 9 cups of flour to 3 cups of water. For every 3 cups of flour, you need 1 cup of water. The ratio 3 : 1 is maintained.

(continued on next page)

Talk About It!
Would this ratio be maintained if you wanted to make half a batch of dough? Explain.

Your Notes

The ratio of cups of flour to cups of water is maintained regardless of how many batches of pizza dough you make. In each batch, there are 3 cups of flour for every 1 cup of water.

Two quantities are in a **proportional relationship** if the two quantities vary and have a constant ratio between them. For example, if a recipe calls for 2 cups of flour for every 1 cup of sugar, the ingredients are in a proportional relationship because, while the number of cups of flour or sugar can vary, the ratio of cups of flour to sugar is constant, 2 : 1.

Some relationships are not proportional relationships. In these cases, a ratio is not maintained. For example, suppose that Pedro is 14 years old and his little brother is 7 years old. The ratio between their current ages is 14 : 7. Pedro is currently twice as old as his brother. Will he always be twice as old?

The bar diagram represents this relationship. Because Pedro is currently twice as old as his brother, the bar diagram representing Pedro's age has twice as many sections as the bar diagram representing his brother's age.

Current Age

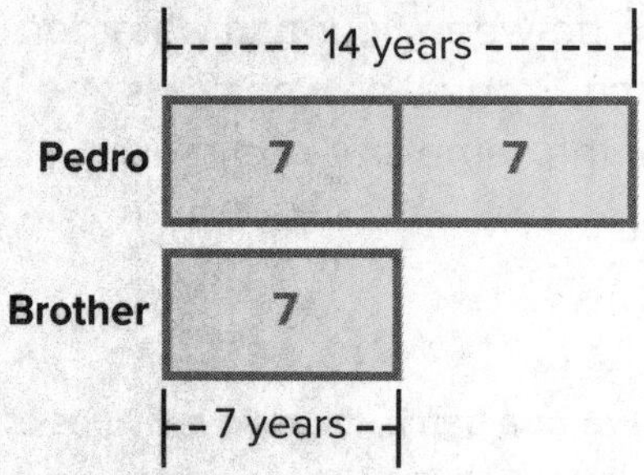

The ratio between Pedro's age and his brother's age is 14 : 7. This ratio is equivalent to 2 : 1 because 14 is twice as great as 7.

In five years, Pedro will be 14 + 5, or 19 years old and his brother will be 7 + 5, or 12 years old. Add a section to each bar diagram to represent the additional 5 years.

Age in Five Years

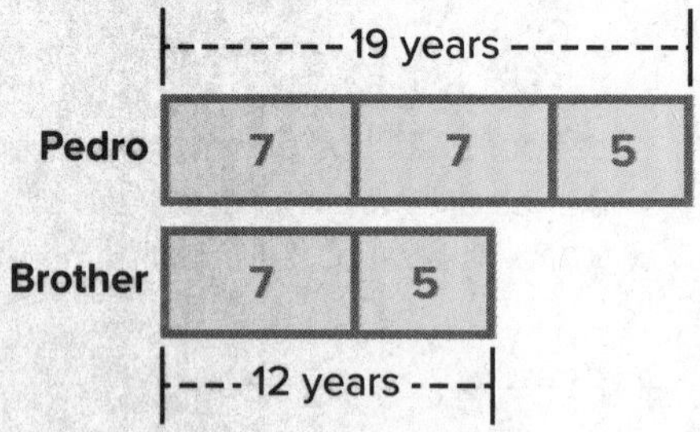

The ratio between Pedro's age and his brother's age in 5 years is 19 : 12.

The bar diagram representing their ages in 5 years shows that the ratio 19 : 12 is not equivalent to 14 : 7 or 2 : 1. In 5 years, Pedro will not be twice as old as his brother because $19 \neq 2(12)$. Because Pedro will not always be twice as old as his brother and the ratio 14 : 7 or 2 : 1 is not maintained, the relationship is not proportional.

Talk About It!

Will there ever be an age, other than 14 and 7, where Pedro is twice as old as his brother? Explain.

Example 1 Identify Proportional Relationships

The recipe for a homemade glass cleaner indicates to use a ratio of 1 part vinegar to 4 parts water. Elyse used 3 tablespoons of vinegar and 12 tablespoons of water to make the cleaner.

Is the relationship between the vinegar and water in the recipe and the vinegar and water in Elyse's cleaning solution a proportional relationship? Explain.

To determine if the relationship is proportional, Elyse must maintain the ratio of vinegar to water. Draw a bar diagram to represent the ratio of ingredients in the recipe.

For every 1 part of vinegar, there are 4 parts of water. The units representing the part do not matter. The units can be cups, quarts, gallons, etc.

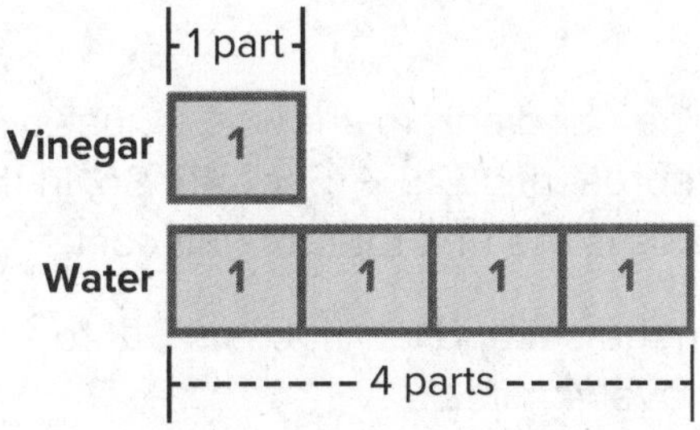

Draw a bar diagram to represent the ratio of ingredients in Elyse's glass cleaner.

Elyse used 3 tablespoons of vinegar, which is 1 part. She used 12 tablespoons of water. Because there are four parts of 3 in 12, she used 4 parts of water.

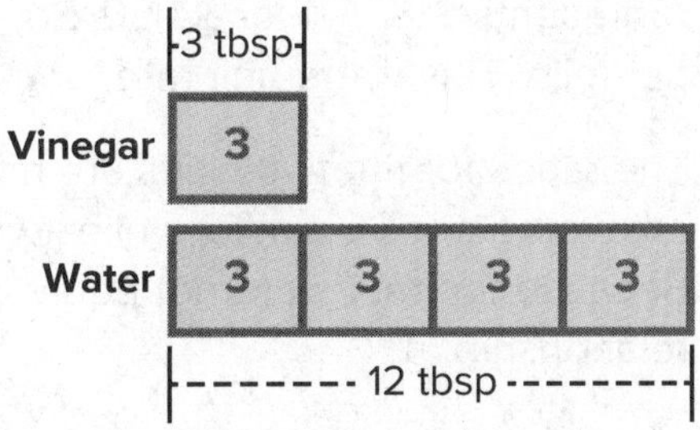

The ratio between vinegar and water was maintained when Elyse used 3 tablespoons of vinegar and 12 tablespoons of water. Because the ratio was maintained, this represents a proportional relationship.

Check

Refer to the recipe for homemade glass cleaner in Example 1. Marcus mixed 1.5 cups of vinegar and 6 cups of water to make his cleaner. Is the relationship between the vinegar and water in the recipe and the vinegar and water in Marcus' cleaning solution a proportional relationship? Explain.

Show your work here

Go Online You can complete an Extra Example online.

Think About It!

What is the relationship between the parts of water and the parts of vinegar in the recipe?

Talk About It!

Would this ratio be maintained if she used 1 cup of vinegar and 4 cups of water? Explain.

Example 2 Identify Proportional Relationships

A 5-mile taxi ride in one city costs Ayana \$25. A 4-mile taxi ride in a different city costs \$18. Assume the cost in each city is the same amount per mile.

Is the relationship between the number of miles and the total cost between the two cities a proportional relationship? Explain.

Draw a diagram to represent the relationship between miles and cost for each city.

In the first city, it costs \$25 to travel 5 miles.

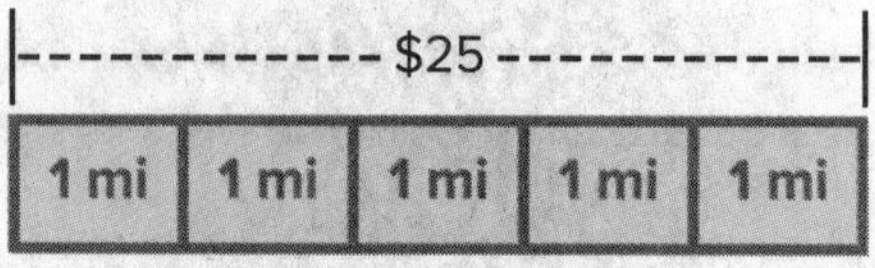

The bar diagram shows 5 equal-size sections. Each section represents \$25 ÷ 5, or \$5. So, in the first city, it costs \$5 per mile. This is the unit rate, or unit cost.

In the second city, it costs \$18 to travel 4 miles.

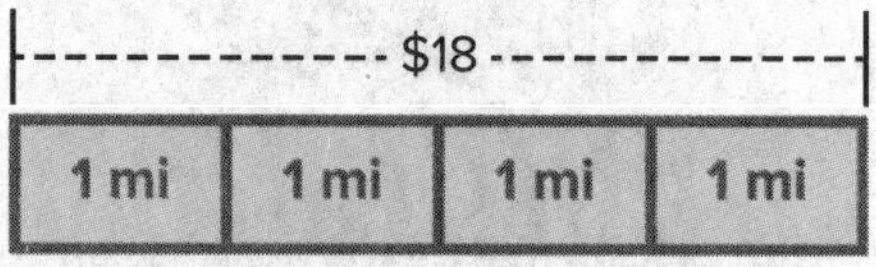

The bar diagram shows 4 equal-size sections. Each section represents \$18 ÷ 4, or \$4.50. So, in the second city, it costs \$4.50 per mile. This is the unit rate, or unit cost.

The ratios for the two cities are not equivalent. In the first city, it costs an average of \$5 a mile, while it costs \$4.50 a mile in the second city. Because the ratios are not equivalent, this is not a proportional relationship.

Talk About It!

What would the cost of a 4-mile taxi ride need to be in the second city so that there was a proportional relationship?

Check

One type of yarn costs \$4 for 100 yards. Another type of yarn costs \$5 for 150 yards. Is the relationship between the number of yards and the cost a proportional relationship between the two types of yarn? Explain.

Show your work here

Go Online You can complete an Extra Example online.

Apply Construction

Christine is building a deck in her backyard. In order to place the posts, she will make concrete using a mixture. The mix requires 1 part water to 2 parts cement to 3 parts sand. The relationship between water, cement, and sand is proportional. If she has 25 pounds of cement and will use it all, how many pounds of sand will she need? One gallon of water weighs about 8.34 pounds. How many gallons of water will she need? Round to the nearest tenth.

1 What is the task?

Make sure you understand exactly what question to answer or problem to solve. You may want to read the problem three times. Discuss these questions with a partner.

First Time Describe the context of the problem, in your own words.
Second Time What mathematics do you see in the problem?
Third Time What are you wondering about?

2 How can you approach the task? What strategies can you use?

Record your observations here

3 What is your solution?

Use your strategy to solve the problem.

Show your work here

4 How can you show your solution is reasonable?

Write About It! Write an argument that can be used to defend your solution.

Talk About It!
Compare and contrast your method for solving this problem with a classmate's method.

Check

A basic slime recipe calls for 1 part borax, 24 parts white glue, and 48 parts water. The relationship between borax, white glue, and water is proportional. If Catalina has 3 tablespoons of borax and will use it all, how many cups of white glue will she need? (*Hint*: 1 cup equals 16 tablespoons)

Show your work here

Go Online You can complete an Extra Example online.

Pause and Reflect

Have you ever wondered when you might use the concepts you learn in math class? What are some everyday scenarios in which you might use what you learned today?

Record your observations here

Name ______________________________ Period ______________ Date ______________

Practice

Go Online You can complete your homework online.

Determine if each situation represents a proportional relationship. Explain your reasoning. (Examples 1 and 2)

1. A salad dressing calls for 3 parts oil and 1 part vinegar. Manuela uses 2 tablespoons of vinegar and 6 tablespoons of oil to make her salad dressing.

2. A specific shade of orange paint calls for 2 parts yellow and 3 parts red. Catie uses 3 cups of yellow paint and 4 cups of red paint to make orange paint.

3. A saltwater solution for an aquarium calls for 35 parts salt to 1000 parts water. Tareq used 7 tablespoons of salt and 200 tablespoons of water.

4. A conveyor belt moves at a constant rate of 12 feet in 3 seconds. A second conveyor belt moves 16 feet in 4 seconds.

5. A tectonic plate in Earth's crust moves at a constant rate of 4 centimeters per year. In a different part of the world, another tectonic plate moves at a constant rate of 30 centimeters in ten years.

6. A strand of hair grows at a constant rate of $\frac{1}{2}$ inch per month. A different strand of hair grows at a constant rate of 4 inches per year.

Test Practice

7. Multiselect One blend of garden soil is 1 part minerals, 1 part peat moss, and 2 parts compost. Select all of the mixtures below that are in a proportional relationship with this blend.

- ☐ 5 ft^3 minerals, 5 ft^3 peat moss, 10 ft^3 compost
- ☐ 10 ft^3 minerals, 15 ft^3 peat moss, 15 ft^3 compost
- ☐ 12 ft^3 minerals, 12 ft^3 peat moss, 24 ft^3 compost
- ☐ 20 ft^3 minerals, 20 ft^3 peat moss, 40 ft^3 compost
- ☐ 100 ft^3 minerals, 100 ft^3 peat moss, 200 ft^3 compost
- ☐ 50 ft^3 minerals, 50 ft^3 peat moss, 50 ft^3 compost

Apply

8. Melanie is making lemonade and finds a recipe that calls for 1 part lemon juice, 2 parts sugar, and 8 parts water. She juices 2 lemons to obtain 6 tablespoons of lemon juice. How much sugar and water will she need to make lemonade with the same ratio of ingredients as the recipe? (*Hint*: 1 cup equals 16 tablespoons)

9. The pizza dough recipe shown makes one batch of dough. Charlie wants to make a half batch. She has 1 cup of flour. How much more flour does she need?

Ingredient	Amount
Flour	3 c
Salt	$\frac{1}{2}$ tsp
Yeast	2 tsp
Water	1 c
Olive Oil	4 tsp

10. MP **Identify Structure** Half of an orange juice mixture is orange concentrate. Explain why the ratio of orange concentrate to water is 1 : 1.

11. MP **Find the Error** One cleaning solution uses 1 part vinegar with 2 parts water. Another cleaning solution uses 2 parts vinegar with 3 parts water. A student says that this represents a proportional relationship because, in each solution, there is one more part of water than vinegar. Find the error and correct it.

12. Patrick made a simple sugar solution using 3 parts sugar and 4 parts water. Thomas made a sugar solution using 6 parts sugar and 7 parts water. Whose solution was more sugary? Explain.

13. MP **Be Precise** How can you use a unit rate to determine if a relationship is proportional?

Tables of Proportional Relationships

I Can... determine whether two quantities shown in a table are in a proportional relationship by testing for equivalent ratios.

What Vocabulary Will You Learn?
constant of proportionality
nonproportional
proportional

Explore Ratios in Tables

Online Activity You will explore how to determine if the ratios between two quantities are equivalent.

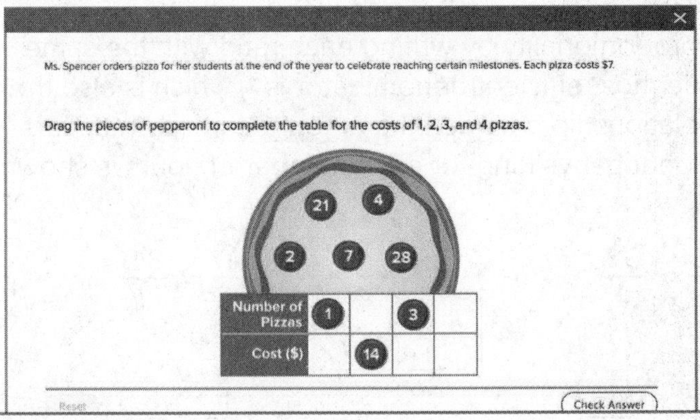

Learn Proportional Relationships and Tables

Two quantities are **proportional** if the ratios comparing them are equivalent.

In the table, all of the ratios comparing the cost to the number of pizzas are equivalent and have a unit rate of $\frac{\$7}{1 \text{ pizza}}$. So, the cost of the order is proportional to the number of pizzas ordered.

Number of Pizzas	1	2	3	4
Cost ($)	7	14	21	28
Cost per Pizza ($)	7	7	7	7

In relationships where these ratios are not equivalent, the two quantities are **nonproportional**.

In the table below, the ratios comparing the cost to the number of pizzas are different. So, the relationship represented by this table is nonproportional.

Number of Pizzas	1	2	3	4
Cost ($)	9	16	23	30
Cost per Pizza ($)	9	8	7.67	7.50

Talk About It!
How can you use ratios to determine if a relationship is proportional?

Your Notes

Think About It!

What makes a relationship proportional?

Talk About It!

Think of a way that Carrie could be paid so the amount she made was not proportional to the number of hours she worked. Explain your reasoning.

Example 1 Proportional Relationships and Tables

Carrie earns $8.50 per hour babysitting.

Is the amount of money she earns proportional to the number of hours she spends babysitting?

Complete the table with the amount of money she earns for babysitting 1, 2, 3, and 4 hours.

Number of Hours	1	2	3	4
Amount Earned ($)				

You can check proportionality by writing each ratio with the same denominator. The most efficient denominator is 1, which is also the unit ratio. The relationship between the amount earned, in dollars, and the hours spent babysitting for each number of hours is shown.

$$\frac{8.5}{1} = \frac{8.5}{1} \qquad \frac{17}{2} = \frac{8.5}{1}$$

$$\frac{25.5}{3} = \frac{8.5}{1} \qquad \frac{34}{4} = \frac{8.5}{1}$$

So, the relationship is proportional because the ratios between the quantities are equivalent and have a unit rate of $8.50 per hour.

Check

An adult elephant drinks about 225 liters of water each day. Is the amount of water proportional to the number of days that have passed? Use the table provided to help answer the question.

Time (days)	1	2	3	4
Water (L)	225	450	675	900

Show your work here

Go Online You can complete an Extra Example online.

Example 2 Proportional Relationships and Tables

A ticket agency charges a $6 service fee on any order. Each ticket for a concert costs $25.

Is the cost of an order proportional to the number of tickets ordered?

The table shows the relationship between the total cost of an order and the number of tickets ordered.

Number of Tickets	1	2	3	4
Total Cost of Order ($)	31	56	81	106

The cost of each additional ticket is an additional $25 because the service fee is only charged once per order.

For each number of tickets, write the relationship of the total cost to the number of tickets as a ratio with a denominator of 1. The first two are done for you.

$\frac{31}{1} = \frac{31}{1}$ $\frac{56}{2} = \frac{28}{1}$

$\frac{81}{3} = \frac{\square}{1}$ $\frac{106}{4} = \frac{\square}{1}$

Because the ratios between the quantities are not the same, the cost of an order is not proportional to the number of tickets ordered.

Check

The table shows how long it took Maria to run laps around the school track. Is the number of laps she ran proportional to the time it took her? Explain.

Laps	2	4	6
Time (s)	150	320	580

Show your work here

Think About It!
How would you begin solving the problem?

Talk About It!
What part of the scenario made the situation nonproportional? Explain your reasoning.

Go Online You can complete an Extra Example online.

Talk About It!

What do you notice about the constant of proportionality and the unit rate?

Think About It!

How are the constant of proportionality and the unit rate related?

Talk About It!

If the relationship was not proportional, would there be a constant of proportionality? Explain.

Learn Identify the Constant of Proportionality

You have learned that two quantities are proportional if the ratios comparing them are equivalent or constant. The constant ratio is called the **constant of proportionality**. The constant of proportionality has the same value as the unit rate.

Creators of a stop motion animation can film 24 frames per second. The table shows the number of frames captured over 1, 2, 3, and 4 seconds.

Number of Seconds	1	2	3	4
Number of Frames	24	48	72	96

What is the constant of proportionality? ________

What is the unit rate? ____________________

Because the constant ratio is $\frac{24}{1}$, the constant of proportionality is 24, and the unit rate is 24 frames per second.

Example 3 Identify the Constant of Proportionality

The winner of a jump rope competition jumped 124 times in 20 seconds and 186 times in 30 seconds.

What is the constant of proportionality?

Write the equivalent ratios so that each ratio has a denominator of 1.

$$\frac{124 \text{ jumps}}{\square} = \frac{\square}{1 \text{ second}}$$

$$\frac{\square}{30 \text{ seconds}} = \frac{\square}{1 \text{ second}}$$

Because the constant ratio is $\frac{6.2}{1}$, the constant of proportionality is 6.2, and the unit rate is 6.2 jumps per second.

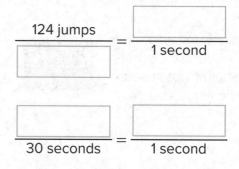

Check

The cost of a birthday party at a skating rink is proportional to the number of guests. The skating rink charges $82.50 for 10 guests.

The choices show the number of guests and the total cost for four different parties that were hosted at different locations. Select each party that was hosted with the same constant of proportionality, or unit rate, as the skating rink's unit rate.

- ☐ 12 guests for $99.00
- ☐ 8 guests for $52.00
- ☐ 15 guests for $97.50
- ☐ 6 guests for $49.50

Show your work here

Go Online You can complete an Extra Example online.

Example 4 Identify the Constant of Proportionality

After a volcanic eruption, lava flows down the slopes of the volcano. The distance the lava flows is proportional to the time.

Time (min)	Distance (m)
5	9
10	18
15	27
20	36

What is the constant of proportionality of the flow of lava?

For each time, find the ratio $\frac{\text{number of meters}}{\text{number of seconds}}$ and rewrite it with a denominator of 1.

$\frac{9}{5} = \frac{\square}{1}$ $\frac{18}{10} = \frac{\square}{1}$

$\frac{27}{15} = \frac{\square}{1}$ $\frac{36}{20} = \frac{\square}{1}$

Because the constant ratios are $\frac{1.8}{1}$, the constant of proportionality is 1.8, and the unit rate is 1.8 meters per second.

Think About It!

Why is a table a good way to organize the information?

Talk About It!

What does the constant of proportionality, 1.8, mean in the context of the problem?

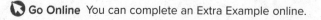

Math History Minute

Erika Tatiana Camacho (1974-) is a Mexican-American mathematical biologist and Associate Professor at Arizona State University. In 2014, she won the Presidential Award for Excellence in Science, Mathematics, and Engineering Mentoring. Her high school teacher and mentor was Jaime Escalante, the subject of the 1988 movie *Stand and Deliver*.

Check

The tables show the amount that three friends earn during a bake sale. Write the correct unit rate that each friend earned in the spaces provided.

Allie

Earnings ($)	17.00	34.00	51.00
Time (h)	1	2	3

Ben

Earnings ($)	9.00	18.00	27.00
Time (h)	0.75	1.5	2.25

Sri

Earnings ($)	40.00	80.00	120.00
Time (h)	2.5	5	7.5

Allie	Ben	Sri

Show your work here

Go Online You can complete an Extra Example online.

Apply Sales Tax

Jalen went shopping for school clothes. The table shows the sales tax for various purchase amounts. Is the sales tax proportional to the purchase amount? What is the total cost, in dollars, for a purchase amount of $84?

Purchase Amount ($)	12	24	36	48
Sales Tax ($)	0.60	1.20	1.80	2.40

1 What is the task?

Make sure you understand exactly what question to answer or problem to solve. You may want to read the problem three times. Discuss these questions with a partner.

First Time Describe the context of the problem, in your own words.
Second Time What mathematics do you see in the problem?
Third Time What are you wondering about?

2 How can you approach the task? What strategies can you use?

Record your observations here

3 What is your solution?

Use your strategy to solve the problem.

Show your work here

4 How can you show your solution is reasonable?

Write About It! Write an argument that can be used to defend your solution.

Talk About It!

If the relationship was not proportional, could you solve this problem? Explain.

Check

The table shows three membership options at a fitness center.

Membership	Cost
Basic	\$20 per class
Fit Plus	\$60 per month plus \$10 per class
Fit Extreme	\$75 per month plus \$30 enrollment fee

Logan chooses the membership that represents a proportional relationship between the number of classes and the monthly cost. Which membership did Logan choose? How much will he spend if he takes 12 classes in a month?

Show your work here

Go Online You can complete an Extra Example online.

Foldables It's time to update your Foldable, located in the Module Review, based on what you learned in this lesson. If you haven't already assembled your Foldable, you can find the instructions on page FL1.

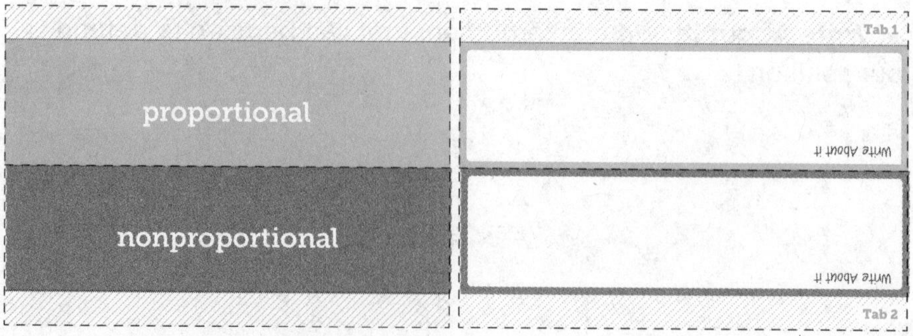

Name ______________________ Period __________ Date __________

Practice

Go Online You can complete your homework online.

For each situation, complete the table given. Does the situation represent a proportional relationship? Explain.

1. The cost of a school lunch is $2.50. **(Example 1)**

Lunches Bought	1	2	3	4
Total Cost ($)				

2. Anna walks her dog at a constant rate of 12 blocks in 8 minutes. **(Example 1)**

Number of Blocks	12	24	36	48
Number of Minutes				

3. Fun Center rents popcorn machines for $20 per hour. In addition to the hourly charge, there is a rental fee of $35. **(Example 2)**

Hours	1	2	3	4
Cost ($)				

4. Jean has $280 in her savings account. Starting next week, she will deposit $30 in her account every week. **(Example 2)**

Weeks	1	2	3	4
Savings ($)				

5. Rocko paid $12.50 for 25 game tickets. Louisa paid $17.50 for 35 game tickets. What is the constant of proportionality? **(Example 3)**

6. A baker, in 70 minutes, iced 40 cupcakes and, in 49 minutes, iced 28 cupcakes. What is the constant of proportionality? **(Example 3)**

7. The table shows the amount of dietary fiber in bananas. Use the table to find the constant of proportionality. **(Example 4)**

Dietary Fiber (g)	9.3	18.6	27.9	37.2
Bananas	3	6	9	12

Test Practice

8. Open Response The table shows the distance traveled by a runner. Use the table to find the constant of proportionality.

Distance (mi)	4.55	13.65	22.75	31.85
Time (h)	0.5	1.5	2.5	3.5

Apply

9. The table shows the amount a restaurant is donating to a local school based on various dinner bills. Is the amount of the donation proportional to the dinner bill? If so, what would be the donation for a dinner bill of $50? If not, explain.

Donations	
Dinner Bill ($)	**Donation ($)**
25	4.50
30	5.40
35	6.30
40	7.20

10. The table shows the cost to mail various letters based on different weights. Is the cost of mailing a letter proportional to the weight? If so, what would be the cost of mailing a 6-ounce letter? If not, explain.

Mailing Costs	
Weight (oz)	**Cost ($)**
1	0.47
2	0.68
3	0.89
4	1.10

11. There are 8 fluid ounces in one cup. If you double the amount of fluid ounces, will the amount of cups also double? Write an argument that can be used to defend your solution.

12. Determine whether the cost of renting equipment is *sometimes*, *always*, or *never* proportional. Explain.

13. MP **Justify Conclusions** Noah ran laps around the school building. The table shows his times. He thinks the number of laps is proportional to his time. Explain how Noah may have come to that conclusion.

Laps	5	10	15
Time (min)	4	6	8

14. **Multiple Representations** Represent the proportional relationship $2 for 5 ears of corn and $4 for 10 ears of corn using another representation.

Graphs of Proportional Relationships

I Can... determine if a relationship is proportional by analyzing its graph and explain what the points (0, 0) and (1, *r*) mean on the graph of a proportional relationship.

Explore Proportional Relationships, Tables, and Graphs

Online Activity You will use Web Sketchpad to explore the graphs of proportional and nonproportional linear relationships.

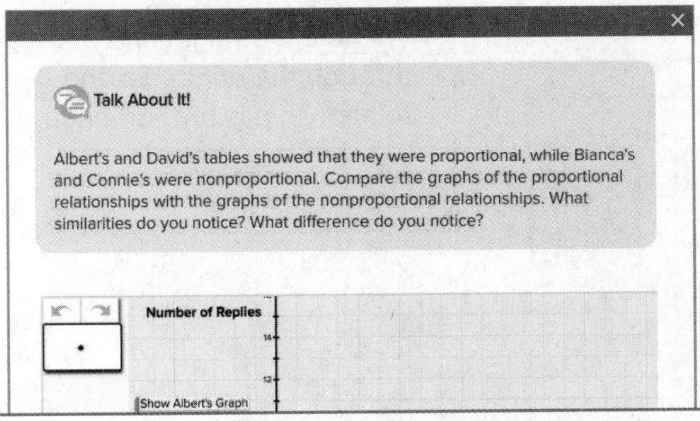

Talk About It!
Albert's and David's tables showed that they were proportional, while Bianca's and Connie's were nonproportional. Compare the graphs of the proportional relationships with the graphs of the nonproportional relationships. What similarities do you notice? What difference do you notice?
Number of Replies
14
12
Show Albert's Graph

Learn Proportional Relationships and Graphs

A graph shows a proportional relationship if it is a straight line through the origin. A graph shows a nonproportional relationship if it is not a straight line, or is a straight line that does not pass through the origin.

Determine if the relationship shown in each graph is *proportional* or *nonproportional*.

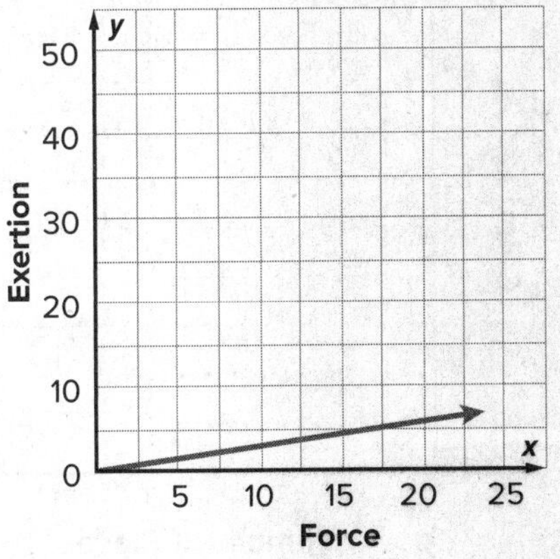

____

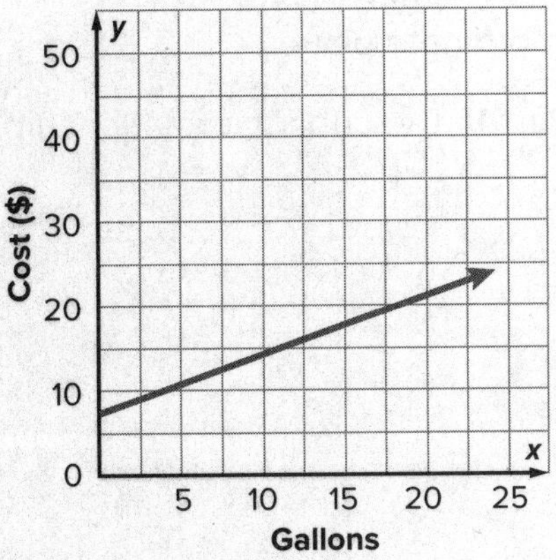

____

Talk About It!

Why does the line of the graph of a proportional relationship need to be straight and pass through the origin?

Your Notes

Think About It!

How will the table help you graph the relationship?

Talk About It!

If the graph of a proportional relationship must be a straight line through the origin, how does the table illustrate that same information?

Example 1 Proportional Relationships and Graphs

A rabbit challenges a tortoise to a race. The table shows the distance that the tortoise moved after 0, 1, 2, and 3 minutes.

Time (min)	Distance (ft)
0	0
1	6
2	12
3	18

Determine whether the number of feet the tortoise moves is proportional to the number of minutes by graphing the relationship on the coordinate plane.

Part A Graph the relationship.

Distance Traveled

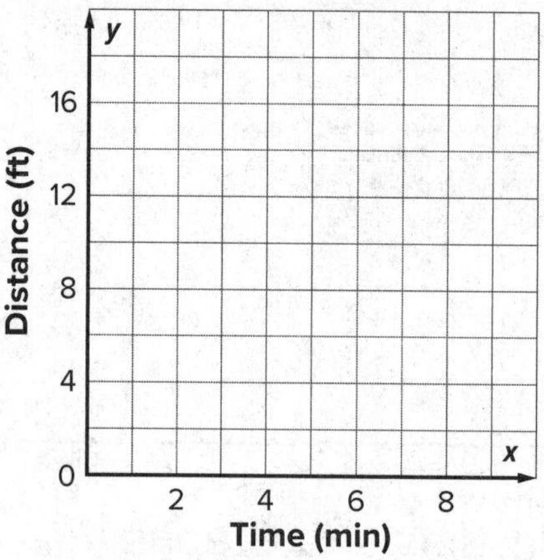

Part B Describe the relationship.

The line is straight and passes through the origin. So, the relationship is proportional.

Check

In a pack of snacks, one piece has 5 Calories, two pieces have 10 Calories, and three pieces have 15 Calories.

Part A Graph the relationship for 0, 1, 2, and 3 pieces on the coordinate plane.

Part B Describe the relationship.

Calories per Piece

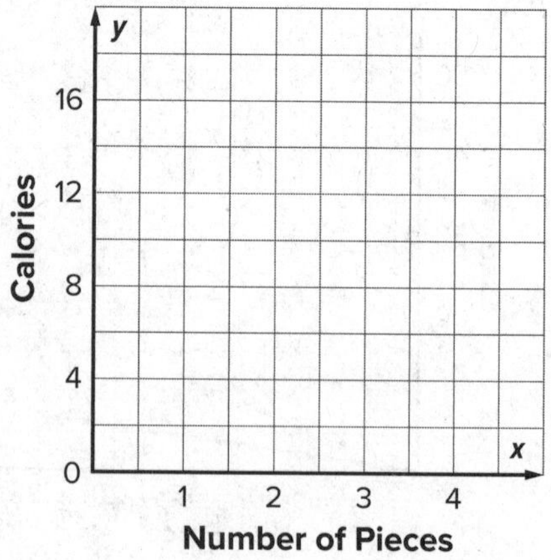

Go Online You can complete an Extra Example online.

Example 2 Proportional Relationships and Graphs

A rabbit challenges a tortoise to a race. The table shows the distance that the rabbit moved after 0, 1, 2, and 3 minutes.

Time (min)	Distance (ft)
0	0
1	8
2	8
3	15

Determine if the distance the rabbit moves is proportional to the time by graphing the relationship on the coordinate plane.

Part A Graph the relationship.

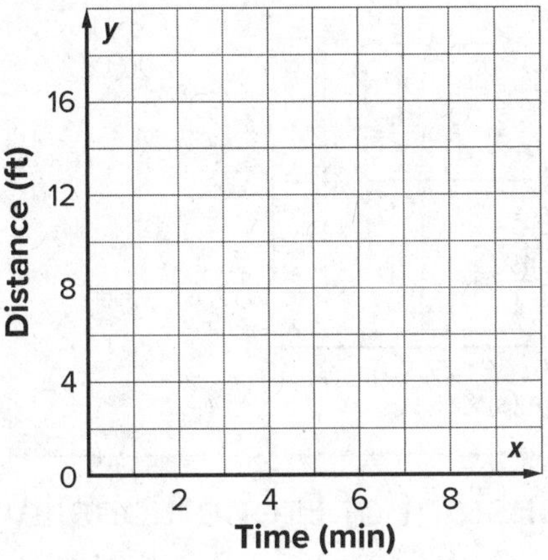

Part B Describe the relationship.

The relationship between time and distance is not proportional because the graph is not a straight line.

Check

The table shows the account balance in a savings account at the end of each week.

Determine if the account balance is proportional to the time by graphing the relationship on the coordinate plane.

Time (wk)	Account Balance ($)
0	10
1	12
2	14
3	16

Part A
Graph the relationship.

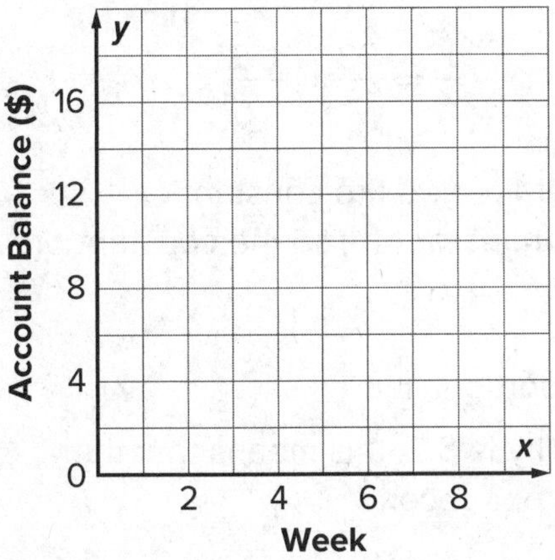

Part B
Describe the relationship.

Talk About It!
Is there another way to determine if the relationship is proportional or nonproportional?

Talk About It!
What part of the scenario makes the relationship nonproportional?

Go Online
You can complete an Extra Example online.

Learn Find the Constant of Proportionality from Graphs

When a proportional relationship is graphed, you can determine the constant of proportionality using any point on the line other than the origin. The constant of proportionality is the ratio of $\frac{y}{x}$ for any point on the line, except (0, 0), when $x = 1$.

Write each ordered pair and write the ratio $\frac{y}{x}$, when $x = 1$, for each.

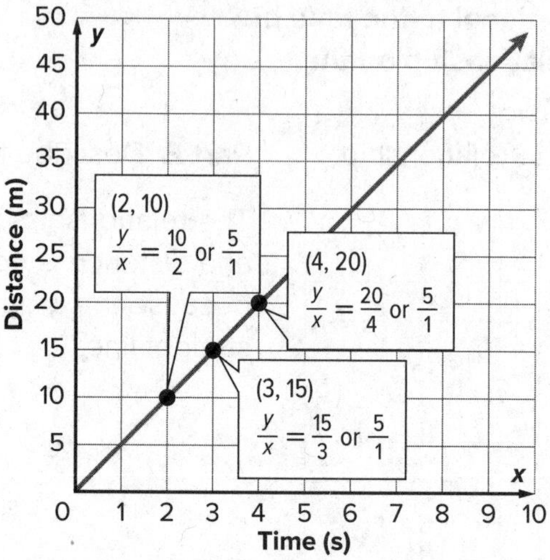

Example 3 Find the Constant of Proportionality from Graphs

In a 100-meter race, assuming the runner's rate is constant, the distance run is proportional to the time spent running. One runner's data are shown on the graph.

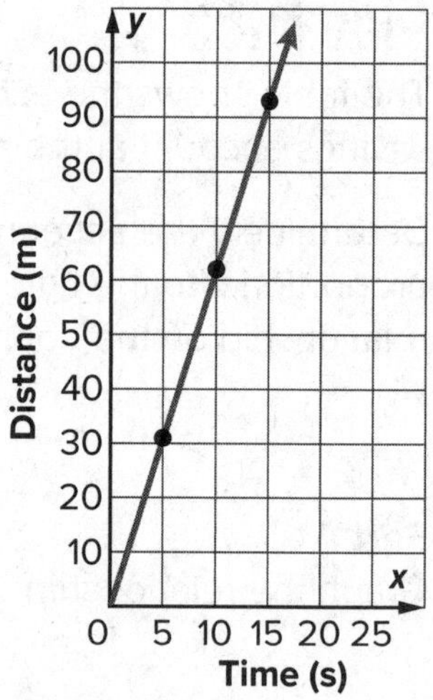

Find the constant of proportionality and describe what it means.

Part A Find the constant of proportionality.

The ratios $\frac{y}{x}$ or $\frac{\text{distance}}{\text{time}}$ for all of the given points are shown.

$\frac{y}{x} = \frac{31}{5} = \frac{6.2}{1}$ $\frac{y}{x} = \frac{62}{10} = \frac{6.2}{1}$ $\frac{y}{x} = \frac{93}{15} = \frac{6.2}{1}$

Write the ratios with a denominator of 1 to find the constant of proportionality. Each ratio has a denominator of 1, so the constant of proportionality is 6.2.

Part B Describe the constant of proportionality.

Because the constant of proportionality is 6.2, this means that the runner moves at a rate of 6.2 meters per second.

Think About It!
How would you begin solving the problem?

Talk About It!
Based on the constant of proportionality, 6.2, how far will the runner travel after 20 seconds? Explain your reasoning.

Check

Briana decides to save money each week for her family vacation. Use the graph to find the constant of proportionality. Then describe what it means.

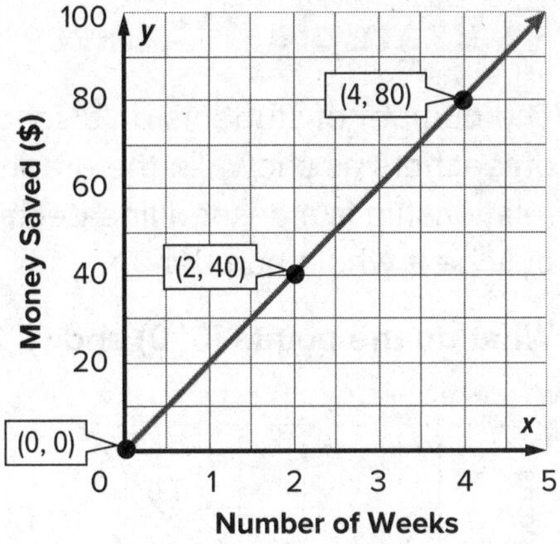

Part A
The constant of proportionality is ________.

Part B
The constant of proportionality means that Briana saves ________ each week.

Go Online You can complete an Extra Example online.

Explore Analyze Points

Online Activity You will explore and analyze the points (0, 0) and (1, *r*) on a graph of a proportional relationship.

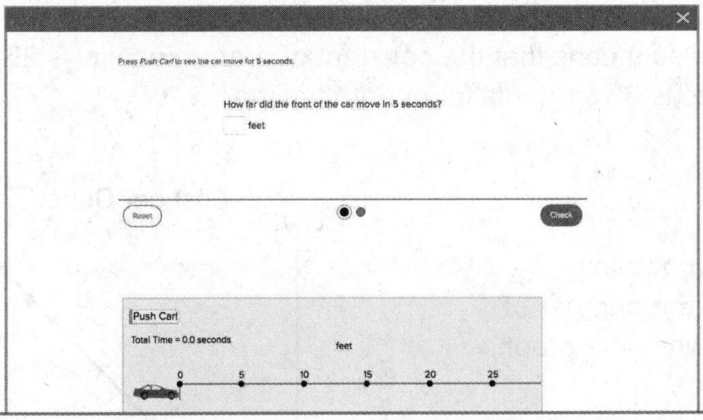

Learn Analyze Points on a Graph

When two quantities are proportional, you can use a graph to find the constant of proportionality and to interpret the point (0, 0).

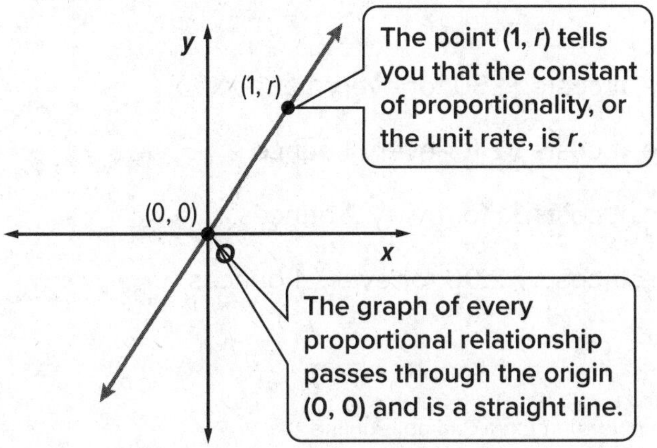

Talk About It!

What is the significance of (0, 0) on the graph of a proportional relationship?

Think About It!

In the ordered pairs, what does the *x*-coordinate represent? the *y*-coordinate?

Talk About It!

Can you find the unit rate from any point on a graph that shows a proportional relationship? Explain your reasoning.

Example 4 Analyze Points on a Graph

The number of students on a school trip is proportional to the number of teachers as shown in the graph. The line representing the relationship is a dashed line because the number of teachers can only be a whole number.

What do the points (0, 0) and (1, 25) represent?

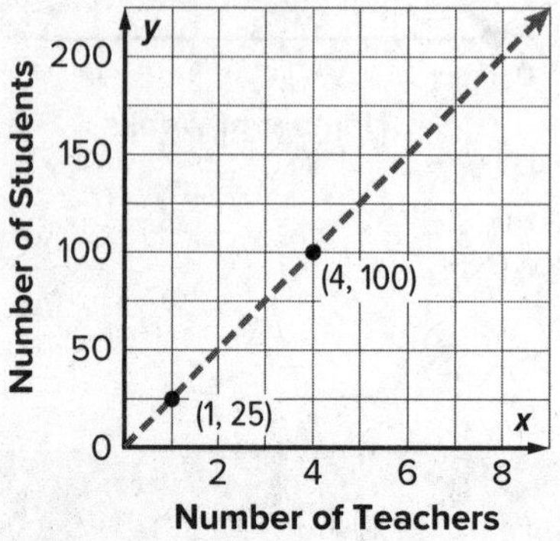

The point (0, 0) means that, for zero teachers, there are zero students.

The point (1, 25) means that, for one teacher, there are __________ students. This also means that the constant of proportionality is 25 and the unit rate is 25 students for every 1 teacher.

Check

The cost of a smoothie is proportional to the number of ounces as shown in the graph.

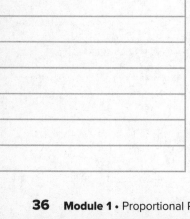

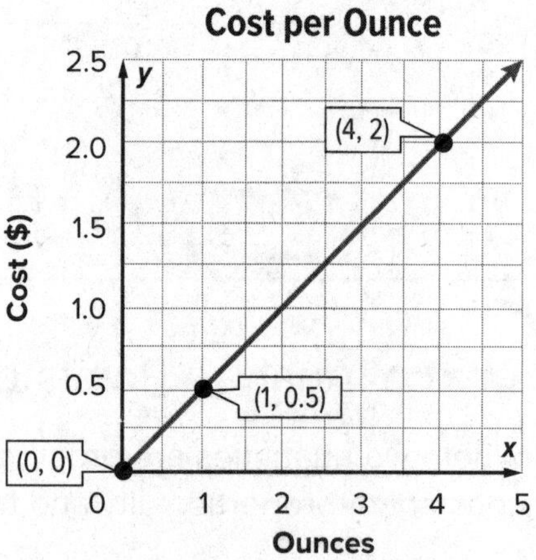

Select all statements that apply.

- ☐ The unit rate is 0.
- ☐ The unit rate is 0.5.
- ☐ The unit rate is 1.
- ☐ For each smoothie, it costs $1.50 for every 1 ounce.
- ☐ For each smoothie, it costs $2 for every 1 ounce.
- ☐ For each smoothie, it costs $1 for every 2 ounces.
- ☐ For each smoothie, it costs $2.00 for every 4 ounces.

Go Online You can complete an Extra Example online.

Apply Fundraising

Michele and Angelo are participating in a school fundraiser. The number of items each student sells after 1, 2, and 3 days is shown in the table. Which ordered pair represents a unit rate for the number of items sold per day?

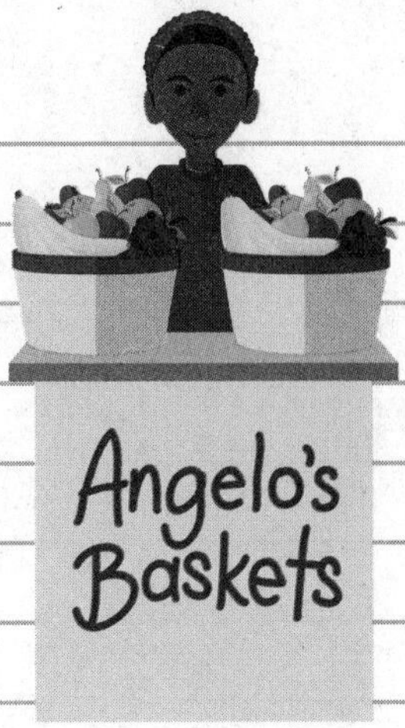

Day	0	1	2	3
Michele's Number of Items	0	3	6	9
Angelo's Number of Items	0	1	4	5

1 What is the task?

Make sure you understand exactly what question to answer or problem to solve. You may want to read the problem three times. Discuss these questions with a partner.

First Time Describe the context of the problem, in your own words.
Second Time What mathematics do you see in the problem?
Third Time What are you wondering about?

2 How can you approach the task? What strategies can you use?

Record your observations here

3 What is your solution?

Use your strategy to solve the problem.

Show your work here

4 How can you show your solution is reasonable?

Write About It! Write an argument that can be used to defend your solution.

Talk About It!

How could you have determined that Angelo's graph was not proportional before graphing the relationship on the coordinate plane?

Check

The heights of two plants are recorded after 1, 2, and 3 weeks. The data are shown in the table. Which ordered pair represents a unit rate for the number of inches grown per week?

Time (weeks)	Plant 1 Height (in.)	Time (weeks)	Plant 2 Height (in.)
0	0	0	0
1	5	1	3
2	6	2	6
3	7	3	9

Go Online You can complete an Extra Example online.

Foldables It's time to update your Foldable, located in the Module Review, based on what you learned in this lesson. If you haven't already assembled your Foldable, you can find the instructions on page FL1.

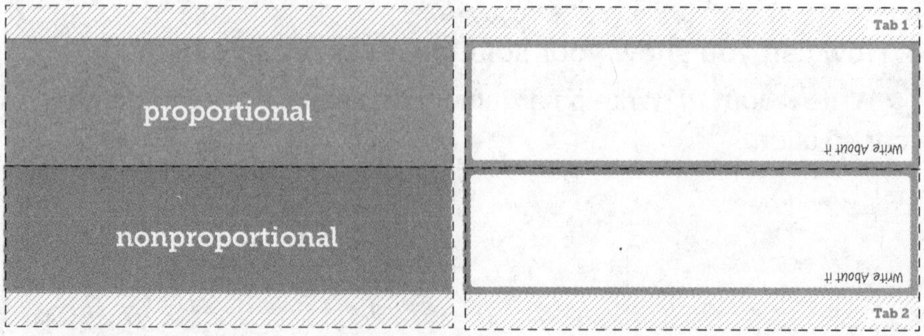

Name ____________________ Period __________ Date __________

Practice

Go Online You can complete your homework online.

1. The cost of pumpkins is shown in the table. Determine whether the cost of a pumpkin is proportional to the number bought by graphing the relationship on the coordinate plane. Explain. **(Example 1)**

Number of Pumpkins	0	1	2	3	4
Cost ($)	0	4	8	12	16

Cost of Pumpkins

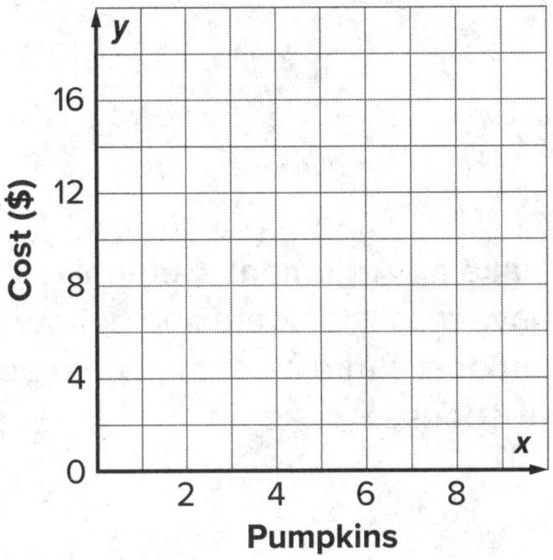

2. The table shows temperatures in degrees Celsius and their equivalent temperatures in degrees Fahrenheit. Determine whether the temperature in degrees Fahrenheit is proportional to the temperature in degrees Celsius by graphing the relationship on the coordinate plane. Explain. **(Example 2)**

Celsius (degrees)	0	5	10	15	20
Fahrenheit (degrees)	32	41	50	59	68

Temperature

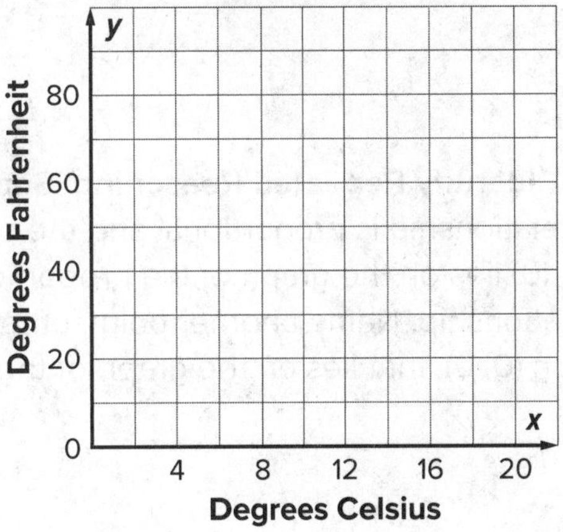

3. The total cost of online streaming is proportional to the number of months. What is the constant of proportionality? **(Example 3)**

Online Streaming of TV Shows/Movies

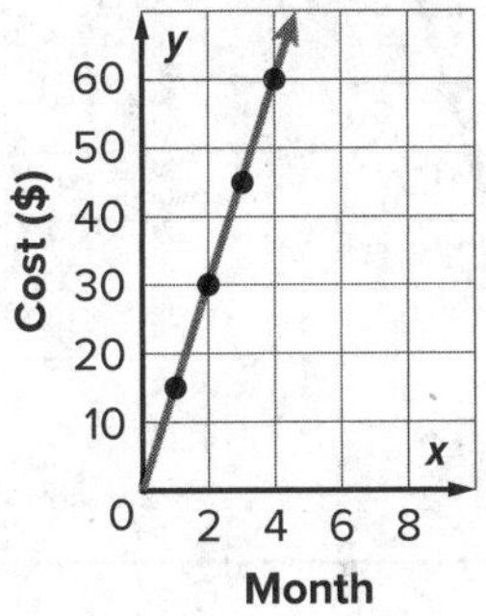

Test Practice

4. Open Response The cost per slice of pizza is proportional to the number of slices as shown in the graph. What do the points (0, 0) and (1, 2) represent? **(Example 4)**

Pizza Slices Cost

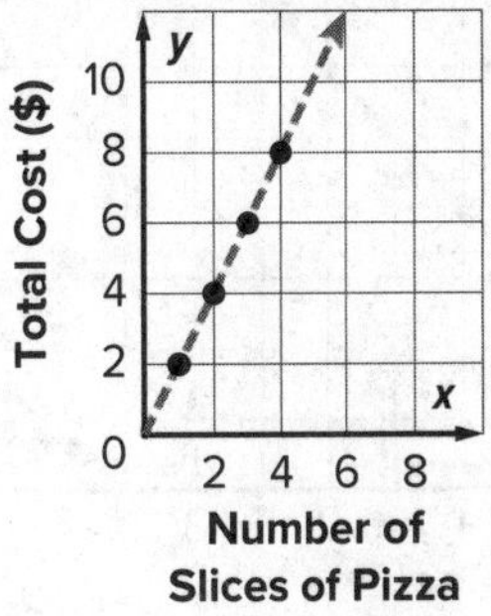

Apply

5. The table shows the number of Calories José and Natalie burned after 1, 2, and 3 minutes of running. Graph the relationship between the number of minutes running and the number of Calories burned for each person. By which ordered pair is a unit rate represented?

	Calories Burned	
Time (min)	José	Natalie
0	0	0
1	8	5
2	15	10
3	23	15

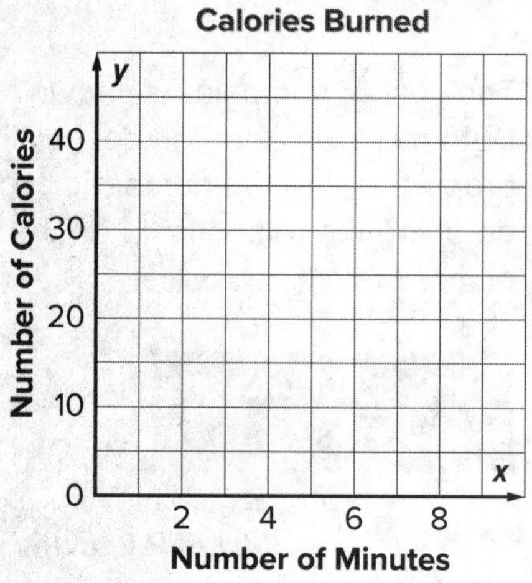

6. MP **Identify Repeated Reasoning** Suppose a relationship is proportional and the point (4, 10) lies on the graph of the proportional relationship. Name another point, other than (0, 0), that lies on the graph of the line.

7. MP **Make an Argument** Determine if a line can have a constant rate and not be proportional. Write an argument to defend your response.

8. MP **Find the Error** Karl said the point (1, 1) represents the constant of proportionality for the graph shown. Find his error and correct it.

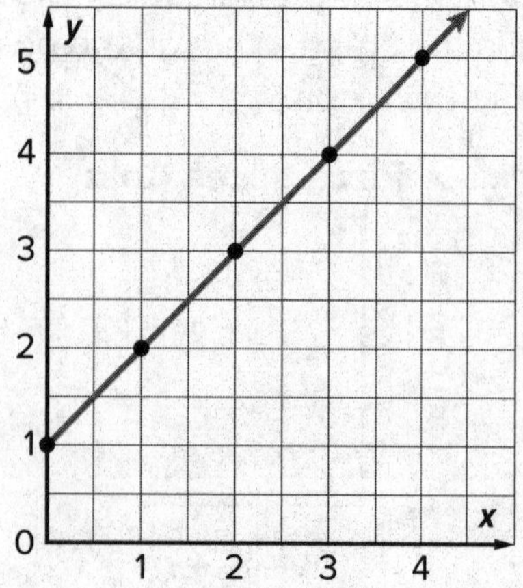

9. **Create** The graph of a proportional relationship is shown. Describe a real-world situation that could be represented by the graph. Be sure to include the meaning of the constant of proportionality.

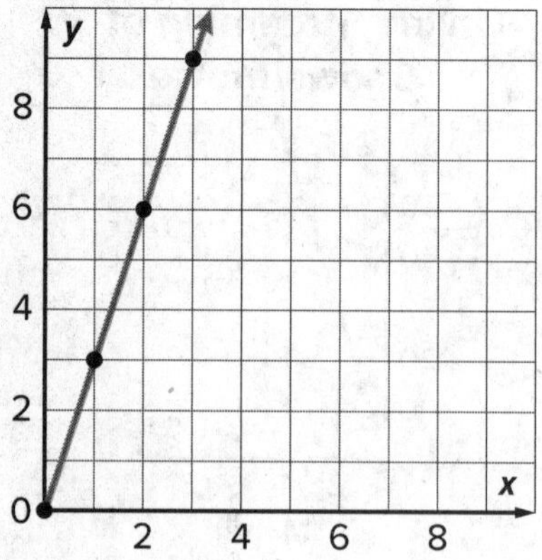

Equations of Proportional Relationships

I Can... write equations to represent proportional relationships and identify the constant of proportionality in the equation representing a proportional relationship.

Explore Proportional Relationships and Equations

Online Activity You will explore the equations of proportional relationships.

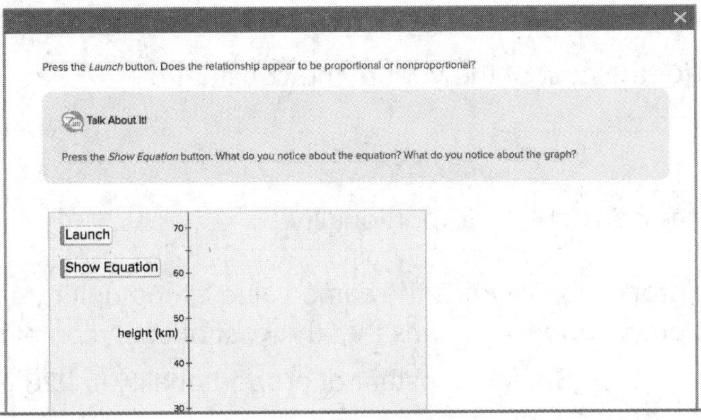

Learn Identify the Constant of Proportionality in Equations

Two quantities are proportional if the ratios comparing them are equivalent. This ratio is called the constant of proportionality. Proportional relationships can be represented by an equation in the form $y = kx$, where k is the constant of proportionality.

Words	Symbols
A linear relationship is proportional when the ratio of y to x is a constant k.	$y = kx$, where $k \neq 0$
Example	**Graph**
$y = 3x$	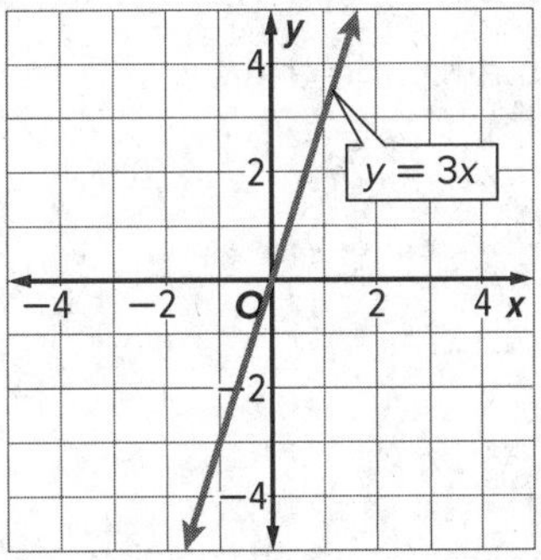

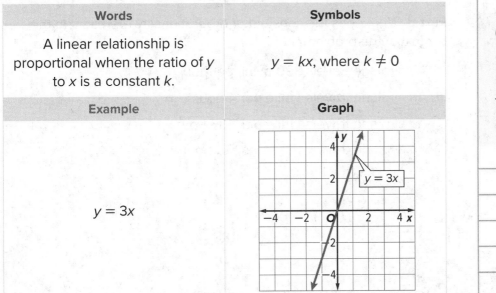

Talk About It!
Can you write an equation in the form $y = kx$ for a line that does not pass through the origin? Explain.

Your Notes

Think About It!

How would you begin solving the problem?

Talk About It!

How much would Olivia pay for 10 yogurts at this same rate?

Example 1 Identify the Constant of Proportionality in Equations

Olivia bought six containers of yogurt for \$7.68. The equation $y = 1.28x$ can be used to represent this situation, where y represents the total cost of the yogurt and x represents the number of containers bought.

Identify the constant of proportionality. Then explain what it represents.

Part A Identify the constant of proportionality.

Compare the two equations.

$y = kx$, where k is the constant of proportionality

$y = 1.28x$

In the equation for the cost of the yogurt, the constant of proportionality k is ________ .

Part B Explain the constant of proportionality.

The constant of proportionality has the same value as the unit rate. The constant of proportionality means that the cost of each container of yogurt is \$ ________. So, the constant of proportionality is 1.28 and the unit rate is \$1.28 per container.

Check

An airplane travels 780 miles in 4 hours. The equation $y = 195x$ models this situation.

Part A What is the constant of proportionality?

Part B What does the constant of proportionality represent in the context of the problem?

Ⓐ An airplane travels 390 miles per hour.

Ⓑ An airplane travels 1 mile per 4 hours.

Ⓒ An airplane travels 1 mile per 585 hours.

Ⓓ An airplane travels 195 miles per hour.

Go Online You can complete an Extra Example online.

Learn Proportional Relationships and Equations

You can use the constant of proportionality, or unit rate, to write an equation in the form $y = kx$ that represents a proportional relationship.

You can find k by writing the ratio comparing y to x. To write the ratio, solve the equation $y = kx$ for k.

$y = kx$	Write the equation.
$\frac{y}{x} = \frac{kx}{x}$	Divide each side by x.
$\frac{y}{x} = k$	Simplify.

The constant of proportionality is the ratio $\frac{y}{x}$, where $x \neq 0$.

Talk About It!

What is the equation for a proportional relationship where the constant of proportionality is 3.25?

Example 2 Proportional Relationships and Equations

Jaycee bought 8 gallons of gas for $31.12.

Write an equation relating the total cost *y* to the number of gallons of gas *x* if it is a proportional relationship.

Step 1 Find the constant of proportionality between cost and gallons.

$$\frac{\text{total cost (\$)}}{\text{number of gallons}} = \frac{31.12}{8}$$

$$= \frac{3.89}{1}$$

The constant of proportionality is ________.

Step 2 Write the equation.

The total cost is $3.89 multiplied by the number of gallons.
Let y = total cost and x = number of gallons.

So, the total cost for any number of gallons can be found using the equation $y = 3.89x$.

Talk About It!

How can you use the equation $y = 3.89x$ to determine the cost for 15 gallons of gas?

Check

A mechanic charges $270 for 5 hours of work on a car. Write an equation that compares the total cost to the number of hours he works on a car if it is a proportional relationship.

Show your work here

Go Online You can complete an Extra Example online.

Example 3 Proportional Relationships and Equations

The perimeter of a square is proportional to the length of one side. A square with 10-inch sides has a perimeter of 40 inches.

Write an equation relating the perimeter of the square to its side length. Then find the perimeter of a square with a 6-inch side.

Part A Write an equation.
The perimeter of a square with 10-inch sides is 40 inches. Write a ratio that compares the perimeter to the side length. Find the constant of proportionality. Find an equivalent ratio with a denominator of 1.

$$\frac{\text{perimeter}}{\text{side length}} = \frac{\square}{\square} = \frac{\square}{1}$$

So, the constant of proportionality is ______.

The perimeter is 4 times the length of a side, so the equation is written $y = 4x$.

Part B Use the equation to find the perimeter of a square with a side length of 6 inches.

$y = 4x$	Write the equation.
$y = 4(6)$	Replace x with 6.
$y = 24$	Multiply.

So, the perimeter of a square with sides of 6 inches is ______ inches.

Think About It!

What do you need to find before you can write the equation?

Talk About It!

Suppose you know the perimeter of a square is 52 inches. Explain how you could use the equation $y = 4x$ to find the length of the side length of the square. Then find the length of the side.

Check

Jack is in charge of the 120,000-gallon community swimming pool. Each spring, he drains the pool in order to clean it. When finished, he refills the pool with fresh water. Jack can fill the pool with 500 gallons of water in 5 minutes.

Part A
Write an equation that represents the relationship between the total number of gallons y and the number of minutes spent filling the pool x.

Part B
How long will it take to become completely filled? ______

Go Online You can complete an Extra Example online.

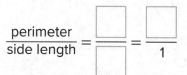

Apply Running

Hugo can run 4 miles in 25 minutes. How many more miles can Hugo run in 90 minutes than in 25 minutes? Assume the relationship is proportional and he runs at a constant rate.

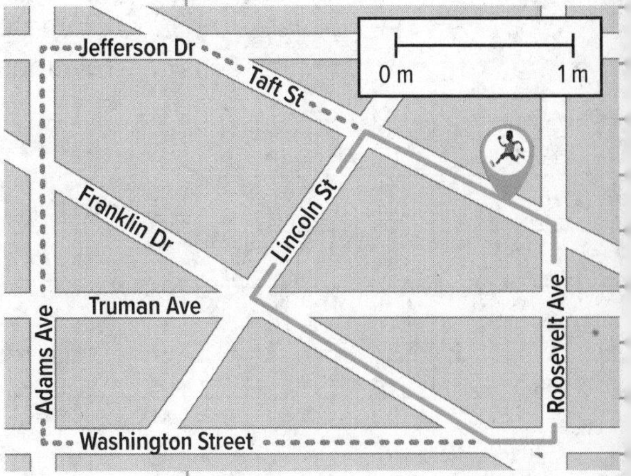

1 What is the task?

Make sure you understand exactly what question to answer or problem to solve. You may want to read the problem three times. Discuss these questions with a partner.

First Time Describe the context of the problem, in your own words.
Second Time What mathematics do you see in the problem?
Third Time What are you wondering about?

2 How can you approach the task? What strategies can you use?

Record your observations here

3 What is your solution?

Use your strategy to solve the problem.

Show your work here

Talk About It!

Can you always assume that Hugo runs at a constant rate for any amount of time predicted? Explain your reasoning.

4 How can you show your solution is reasonable?

Write About It! Write an argument that can be used to defend your solution.

Check

Emily can ride her bike 15 kilometers in 45 minutes. How many more kilometers can she bike in two hours than in 45 minutes? Assume the relationship is proportional and she always bikes at the same rate.

Show your work here

Go Online You can complete an Extra Example online.

Foldables It's time to update your Foldable, located in the Module Review, based on what you learned in this lesson. If you haven't already assembled your Foldable, you can find the instructions on page FL1.

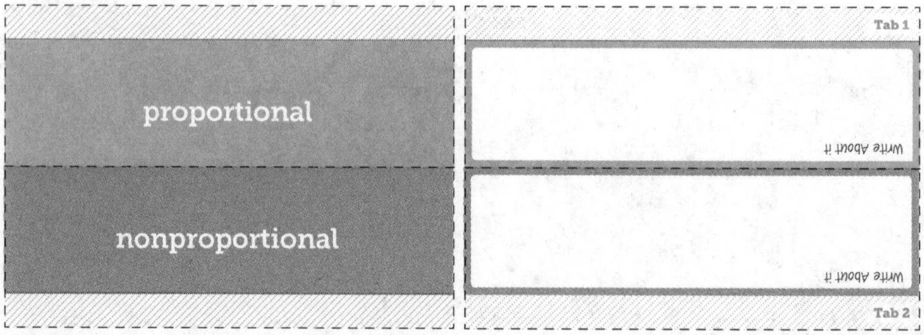

Name ______________________ Period __________ Date __________

Practice

Go Online You can complete your homework online.

1. Liv earns $9.50 for every two bracelets she sells. The equation $y = 4.75x$ can be used to represent this situation. What is the constant of proportionality? What does the constant of proportionality represent in the context of the problem? **(Example 1)**

2. John ran 3 miles in 25.5 minutes. The equation $y = 8.5x$ can be used to represent this situation. What is the constant of proportionality? What does the constant of proportionality represent in the context of the problem? **(Example 1)**

3. Lincoln bought 3 bottles of an energy drink for $4.50. Write an equation relating the total cost y to the number of energy drinks bought x. **(Example 2)**

4. The total cost of renting a cotton candy machine for 4 hours is $72. What equation can be used to model the total cost y for renting the cotton candy machine x hours? **(Example 2)**

5. Marley used 7 cups of water to make 4 loaves of French bread. What equation can be used to model the total cups of water needed y for making x loaves of French bread? How many cups of water do you need for 6 loaves of French bread? **(Example 3)**

6. Mrs. Henderson used $6\frac{3}{4}$ yards of fabric to make 3 elf costumes. What equation can be used to model the total number of yards of fabric y for x costumes? How many yards of fabric do you need for 7 elf costumes? **(Example 3)**

Test Practice

7. **Multiselect** The table shows the cost of 4 movie tickets at two theaters. Select the statements that are true about the situation.

Theater	Cost ($)
Movies Galore	30
Star Cinema	31

- ☐ The equation $y = 7.75x$ models the cost for tickets at Star Cinema.
- ☐ The equation $y = 30x$ models the cost for tickets at Movies Galore.
- ☐ The total cost of 9 tickets at Star Cinema would cost $69.75.
- ☐ The total cost for 1 ticket at Movies Galore is $30.

Apply

8. Roman can type 3 pages in 60 minutes. How many more pages can Roman type in 90 minutes than in 60 minutes? Assume the relationship is proportional and he types at a constant rate.

9. On average, Asia makes 14 out of 20 free throws. Assuming the relationship is proportional, how many more free throws is she likely to make if she shoots 150 free throws?

10. Evan earned $26 for 4 hours of babysitting. What equation can be used to model his total earnings *y* for babysitting *x* hours? Then graph the equation on the coordinate plane. What is the unit rate? How is that represented on the graph?

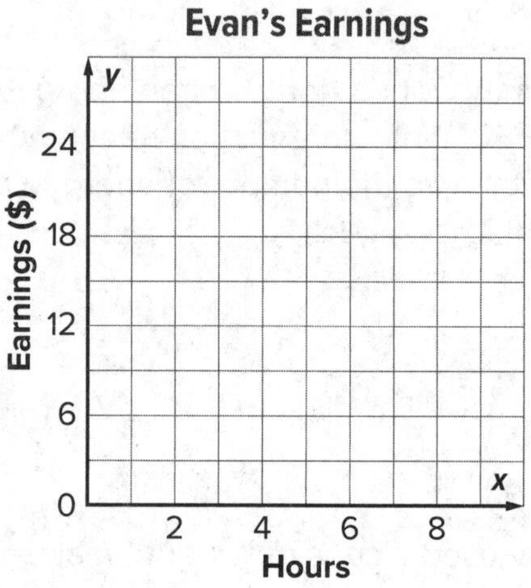

11. MP **Persevere with Problems** The Diaz family spent $38.25 on 3 large pizzas. What is the cost of one large pizza? Assume the situation is proportional. Explain how you solved.

12. MP **Use a Counterexample** Determine whether the statement is *true* or *false*. If false, give a counterexample.

The constant of proportionality in an equation can never be 0.

13. MP **Justify Conclusions** A recipe for homemade modeling clay includes $\frac{1}{3}$ cup of salt for every cup of water. If there are 6 cups of salt, how many gallons of water are needed? Identify the constant of proportionality. Explain your reasoning.

Solve Problems Involving Proportional Relationships

I Can... solve problems involving proportional relationships by making a table, using a graph, or writing an equation.

What Vocabulary Will You Learn?
proportion

Learn Proportions

A **proportion** is an equation stating that two ratios are equivalent. Suppose a recipe indicates a ratio of 3 cups of milk for every 4 cups of flour. The ratio 3 cups of milk to 4 cups of flour is equivalent to 6 cups of milk to 8 cups of flour. This relationship can be written as a proportion.

cups of milk → $\frac{3}{4} = \frac{6}{8}$ ← cups of milk
cups of flour → ← cups of flour

The equals sign means that the ratios are equivalent and the original ratio of 3 cups of milk for every 4 cups of flour is maintained. The unit rate is $\frac{3}{4}$ cup of milk for every cup of flour.

You can use any representation to solve a problem involving a proportional relationship. The multiple representations table shows different methods for finding the number of cups of milk needed if 3 cups of flour are used.

Words

The unit rate is $\frac{3}{4}$ cup of milk for every cup of flour. If you use 3 cups of flour, you need $3\left(\frac{3}{4}\right)$ or $2\frac{1}{4}$ cups of milk.

Ratio Table

÷4 ×3

Cups of Milk	3	$\frac{3}{4}$	$\frac{9}{4}$ or $2\frac{1}{4}$
Cups of Flour	4	1	3

÷4 ×3

Graph

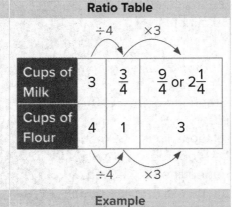

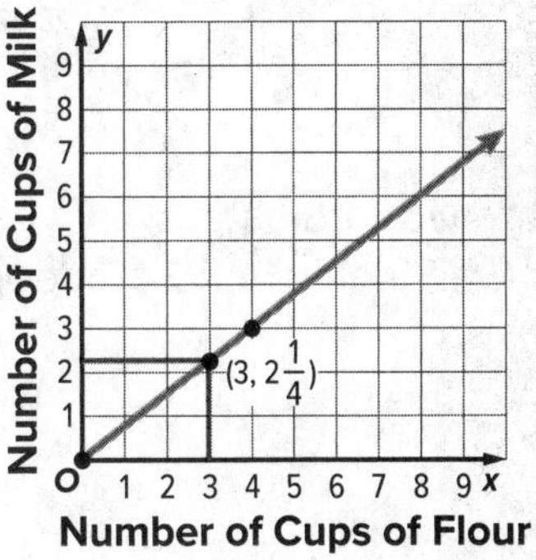

Example

The unit rate is $\frac{3}{4}$ cup of milk to 1 cup of flour, so the constant of proportionality is $\frac{3}{4}$. Let *y* represent the number of cups of milk, and *x* represent the number of cups of flour.

$$y = \frac{3}{4}x$$

$$y = \frac{3}{4}(3), \text{ or } 2\frac{1}{4}$$

Talk About It!

How does each representation show the unit rate?

Your Notes

Think About It!

Is the expected wait time for 240 people less than, greater than, or equal to 40 minutes? How do you know?

Talk About It!

How does the ratio table illustrate the unit rate, or constant of proportionality?

Talk About It!

In this situation, how accurate is the use of a graph when finding the *x*- or *y*-coordinate?

Example 1 Solve Problems Involving Proportional Relationships

The wait time to ride a roller coaster is 20 minutes when 160 people are in line.

How long is the expected wait time when 240 people are in line? Assume the relationship is proportional.

Method 1 Use a table.

There is no whole number by which you can multiply 160 by to obtain 240.

Scale back to find the unit rate. Because $20 \div 20 = 1$, divide 160 by 20 to obtain 8. When 8 people are in line, the expected wait time is 1 minute.

÷20

Time (min)	1	20	?
Number of People	8	160	240

÷20

Then scale forward to find the expected wait time when 240 people are in line. Because $8(30) = 240$, multiply 1 by 30 to obtain 30.

×30

Time (min)	1	20	?
Number of People	8	160	240

×30

When 240 people are in line, the expected wait time is 30 minutes.

Method 2 Use a graph.

Graph the points (0, 0) and (20, 160). Because the relationship is assumed to be proportional, draw a dotted line connecting the points. The line passes through the origin. Determine the corresponding *x*-coordinate for when the *y*-coordinate is 240.

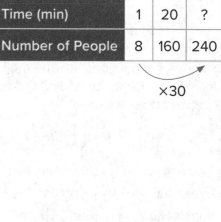

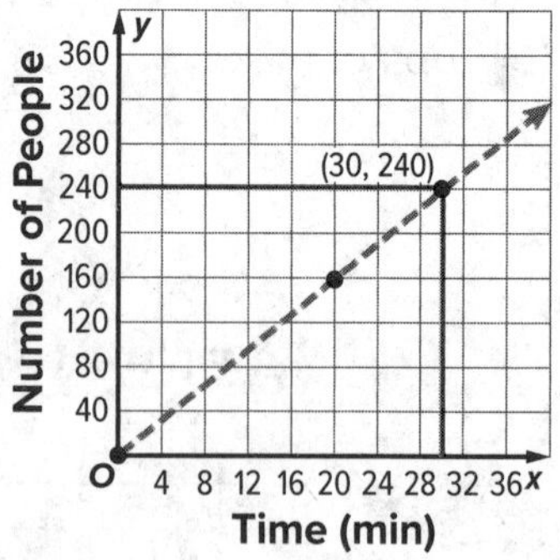

The corresponding *x*-coordinate when the *y*-coordinate is 240 appears to be 30. So, the expected wait time when 240 people are in line is 30 minutes.

(continued on next page)

Method 3 Use an equation.

The equation that represents a proportional relationship is of the form $y = kx$, where k is the constant of proportionality, or unit rate.

Find the unit rate, the number of people in line when the expected wait time is 1 minute. Let y represent the number of people in line and x represent the number of minutes.

$y = kx$	Write the equation.
$160 = k(20)$	Replace y with 160 and x with 20.
$\frac{160}{20} = \frac{k(20)}{20}$	Divide each side by 20 to find the value of k.
$8 = k$	Simplify. The unit rate, or constant of proportionality, is 8.

When 8 people are in line, the expected wait time is 1 minute. Use the equation $y = 8x$ to find the expected wait time when 240 people are in line.

$y = 8x$	Write the equation. The constant of proportionality is 8.
$240 = 8x$	Replace y with 240.
$\frac{240}{8} = \frac{8x}{8}$	Divide each side by 8 to find the value of x.
$30 = x$	Simplify.

So, the expected wait time when 240 people are in line is 30 minutes.

Check

Matthew paid $49.45 for five used video games of equal cost. The relationship between the number of video games and the total cost is proportional. What is the total cost for 11 used video games? Use any strategy.

Show your work here

Talk About It!

Suppose a classmate drew the double number line shown to represent and solve this problem. Is this a valid method? Explain. Does this method show the unit rate? Is the unit rate necessary to solve the problem? Explain.

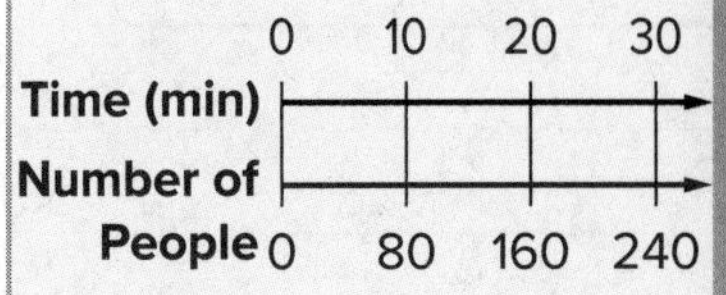

Go Online You can complete an Extra Example online.

Example 2 Solve Problems Involving Proportional Relationships

After two hours, the outside air temperature had risen 6°F. The temperature is forecasted to continue to increase at this same rate for the next several hours.

At this rate, in how many hours will it take the temperature to rise an additional 13°F?

Choose a strategy for solving this problem. For this problem, using an equation is an advantageous strategy because calculations with fractional values will be involved. You know this because there is no whole number by which you can multiply 6 to obtain 13.

The equation that represents a proportional relationship is of the form $y = kx$, where k is the constant of proportionality, or unit rate.

Find the unit rate, the rise in temperature in degrees Fahrenheit per hour. Let y represent the rise in temperature in degrees Fahrenheit and x represent the number of hours.

$y = kx$	Write the equation.
$6 = k(2)$	Replace y with 6 and x with 2.
$\frac{6}{2} = \frac{k(2)}{2}$	Divide each side by 2 to find the value of k.
$3 = k$	Simplify. The unit rate, or constant of proportionality, is 3.

Each hour, the temperature is forecasted to rise 3°F.

Use the equation $y = 3x$ to find the number of hours it is expected to take the temperature to rise an additional 13°F.

$y = 3x$	Write the equation. The constant of proportionality is 3.
$13 = 3x$	Replace y with 13.
$\frac{13}{3} = \frac{3x}{3}$	Divide each side by 3 to find the value of x.
$4\frac{1}{3} = x$	Simplify.

So, after $4\frac{1}{3}$ hours, or 4 hours and 20 minutes, the temperature is forecasted to rise an additional 13°F.

Think About It!

A classmate said that the number of hours for the temperature to rise an additional 13°F is greater than 4 hours. Is this true? How do you know without calculating?

Talk About It!

What other strategies can you choose to solve this problem? Explain why they may or may not be advantageous.

Check

Brooke bought 8 bottled teas for $13.52. Assume the relationship between the number of bottled teas and total cost, in dollars, is proportional. How much can Brooke expect to pay for 12 bottled teas?

Go Online You can complete an Extra Example online.

Apply Blood Drives

At a recent statewide blood drive, the ratio of Type O to non-Type O donors was 37 : 43. Suppose there are 300 donors at a local blood drive. About how many are Type O?

Go Online watch the animation online.

1 What is the task?

Make sure you understand exactly what question to answer or problem to solve. You may want to read the problem three times. Discuss these questions with a partner.

First Time Describe the context of the problem, in your own words.
Second Time What mathematics do you see in the problem?
Third Time What are you wondering about?

2 How can you approach the task? What strategies can you use?

Record your observations here

3 What is your solution?

Use your strategy to solve the problem.

Show your work here

4 How can you show your solution is reasonable?

Write About It! Write an argument that can be used to defend your solution.

Talk About It!
How can you use the ratio 37 : 43 to solve the problem?

Check

The ratio of girls to boys in a school is 2 : 3. The relationship between the number of girls and the number of boys is proportional. How many boys are there if there are 345 students in the school?

Show your work here

Go Online You can complete an Extra Example online.

Pause and Reflect

Compare and contrast the different methods presented in this lesson for solving problems involving proportional relationships.

Record your observations here

Name ____________________ Period __________ Date __________

Practice

Go Online You can complete your homework online.

For each problem, use any method. Assume each relationship is proportional. (Examples 1 and 2)

1. For every three girls taking classes at a martial arts school, there are 4 boys who are taking classes. If there are 236 boys taking classes, predict the number of girls taking classes at the school.

2. A grading machine can grade 96 multiple choice tests in 2 minutes. If a teacher has 300 multiple choice tests to grade, predict the number of minutes it will take the machine to grade the tests.

3. A 6-ounce package of fruit snacks contains 45 pieces. How many pieces would you expect in a 10-ounce package?

4. Of the 50 students in the cafeteria, 7 have red hair. If there are 750 students in the school, predict the number of students who have red hair.

5. The wait times for two different rides are shown in the table. If there are 120 people in line for the swings, how long can you expect to wait to ride the ride?

Ride	Wait Times
Carousel	6 minutes for 48 people in line
Swings	12 minutes for 75 people in line

Test Practice

6. **Open Response** Ingrid types 3 pages in the same amount of time that Tanya types 4.5 pages. If Ingrid and Tanya start typing at the same time and continue at their respective rates, how many pages will Tanya have typed when Ingrid has typed 11 pages?

Apply

7. The ratio of kids to adults at a school festival is 11 : 7. Suppose there are a total of 810 kids and adults at the festival. How many adults are at the festival?

8. The ratio of laptops to tablets in the stock room of a store is 13 : 17. If there are a total of 90 laptops and tablets in the stock room, how many laptops are in the stock room?

9. MP **Persevere with Problems** Lisa is painting the exterior surfaces at her home. A gallon of paint will cover 350 square feet. How many gallons of paint will Lisa need to paint one side of her fence? Explain how you solved.

Item to Paint	Length (ft)	Width (ft)
Fence	26	7
Barn Door	11	6

10. MP **Find the Error** The rate of growth for a plant is 0.2 centimeter per 0.5 day. A student found the number of days for the plant to grow 3.6 centimeters to be 1.44 days. Find the error and correct it.

11. **Create** Write a real-world problem involving a proportional relationship. Then solve the problem.

12. MP **Be Precise** When is it more beneficial to solve a problem involving a proportional relationship using an equation than using a graph?

Review

Foldables Use your Foldable to help review the module.

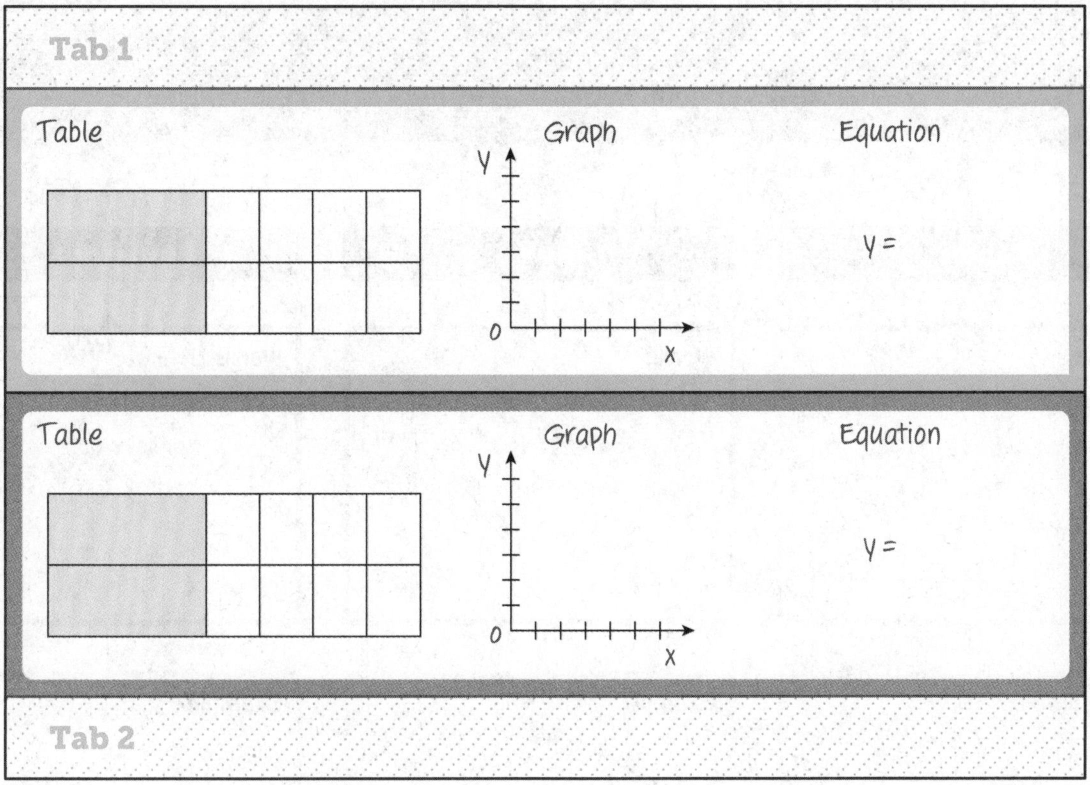

Rate Yourself!

Complete the chart at the beginning of the module by placing a checkmark in each row that corresponds with how much you know about each topic after completing this module.

Write about one thing you learned.

Write about a question you still have.

Reflect on the Module

Use what you learned about proportional relationships to complete the graphic organizer.

Essential Question

What does it mean for two quantities to be in a proportional relationship?

Tables

Words

Example

x				
y				

Counterexample

x				
y				

Graphs

Words

Example

Counterexample

Equations

Words

Example

Counterexample

Name ______________________ Period __________ Date __________

Test Practice

1. Multiple Choice Noreen can walk $\frac{1}{3}$ mile in 12 minutes. What is her average speed in miles per hour? **(Lesson 1)**

Ⓐ 36 miles per hour

Ⓑ 12 miles per hour

Ⓒ $\frac{1}{4}$ mile per hour

Ⓓ $1\frac{2}{3}$ miles per hour

2. Open Response The table shows the amount spent on tomatoes at three different stands at a farmer's market. Which stand sold their tomatoes at the least expensive price per pound, and what is that price? Round to the nearest cent, if necessary. **(Lesson 1)**

Stand	Weight (lb)	Cost ($)
A	$\frac{3}{4}$	2.00
B	$1\frac{1}{4}$	3.45
C	$5\frac{1}{2}$	14.85

3. Multiselect One recipe for homemade playdough calls for 4 parts flour, 1 part salt, and 2 parts water. Select all of the mixtures below that are in a proportional relationship with this recipe. **(Lesson 2)**

☐ 8 cups flour, 2 cups salt, 4 cups water

☐ 2 cups flour, $\frac{1}{2}$ cup salt, $\frac{1}{2}$ cup water

☐ 6 cups flour, $1\frac{1}{2}$ cups salt, 3 cups water

☐ 10 cups flour, 1 cup salt, 2 cups water

4. Open Response The ratio of Braydon's number of laps he ran to the time he ran is 6 : 2. The ratio of Monique's number of laps she ran to the time she ran is 10 : 4. Explain why these ratios are not in a proportional relationship. **(Lesson 2)**

5. Open Response One month, Miko bought two books at a used book store for a total of $2. Over the next few months, she bought four books for a total of $5, six books for a total of $7, and eight books for a total of $10. Is the cost proportional to the number of books purchased? Explain. **(Lesson 3)**

6. Table Item The table shows the time and distance Hector walked in a 5-kilometer-long walk for charity. The distance is proportional to the time walked. Complete the table to show the times of his last three walks. **(Lesson 3)**

Distance (km)	Time (min)
1.5	13.5
2	
4.5	
5	

7. Multiselect The relationship between the number of slices of pizza purchased and the number of students served is shown in the graph. Select all of the statements that are true. **(Lesson 4)**

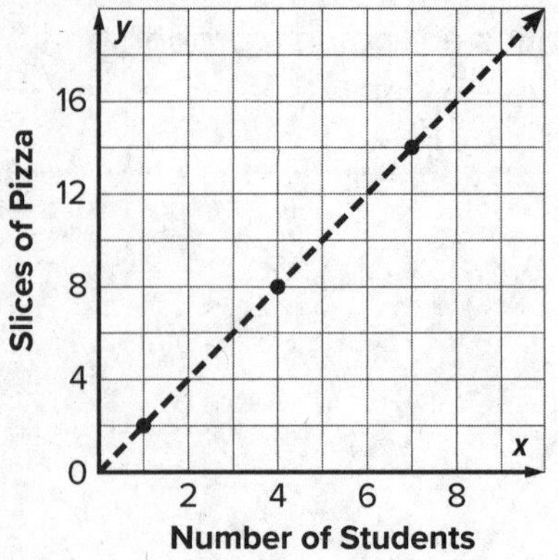

- [ ] The relationship is proportional.
- [ ] The point (9, 18) satisfies this relationship.
- [ ] The constant of proportionality is $\frac{1}{2}$.
- [ ] The constant of proportionality is 2.

8. Grid The cost of dance lessons is $12 for 1 lesson, $22 for 2 lessons, and $32 for 3 lessons. **(Lesson 4)**

A. Graph the ordered pairs on the coordinate plane.

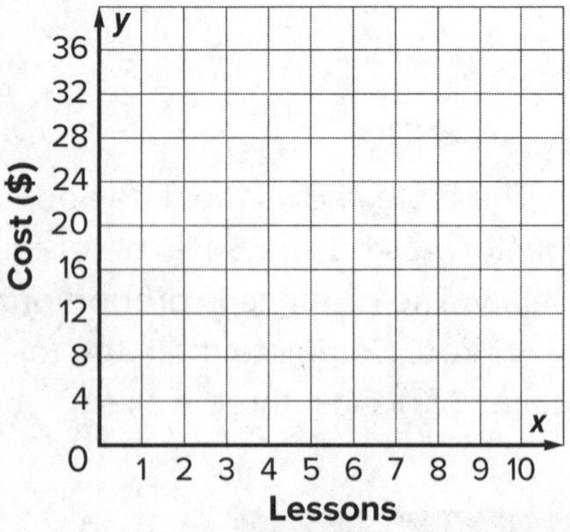

B. Determine whether the cost is proportional to the number of lessons. Explain your reasoning.

9. Equation Editor Mrs. Jameson paid $202.50 for a group of 9 students to visit an amusement park. **(Lesson 5)**

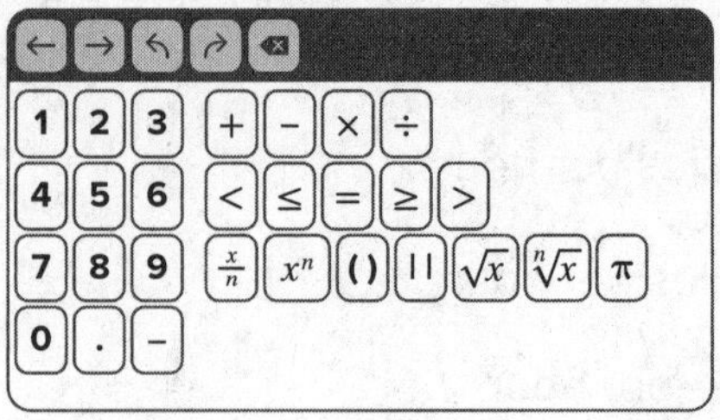

A. Write an equation relating the total cost *y* and the number of students *x* attending the park.

B. What would be the total cost if four more students wanted to join the group?

10. Equation Editor A homeowner whose house was assessed at $120,000 pays $1,800 in taxes. At the same rate, what is the tax on a house assessed at $135,000? **(Lesson 6)**

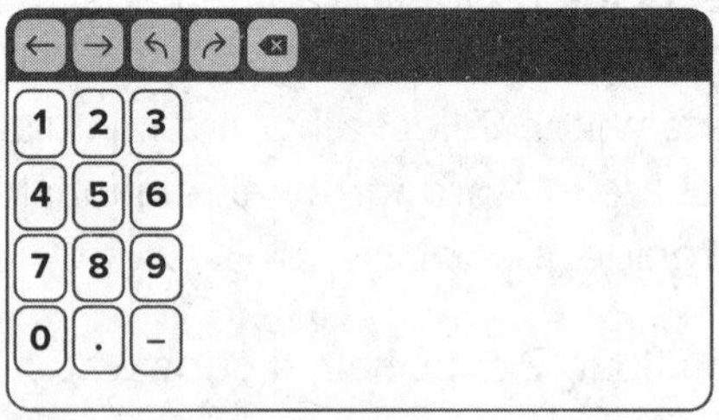

Module 2

Solve Percent Problems

Essential Question

How can percent describe the change of a quantity?

What Will You Learn?

Place a checkmark (✓) in each row that corresponds with how much you already know about each topic **before** starting this module.

KEY

□ — I don't know. ◇ — I've heard of it. ☆ — I know it!

	Before			After		
	□	◇	☆	□	◇	☆
finding percent of change						
solving problems involving taxes						
solving problems involving tips and markups						
solving problems involving discounts						
solving problems involving interest						
solving problems involving commissions and fees						
finding percent error						

Foldables Cut out the Foldable and tape it to the Module Review at the end of the module. You can use the Foldable throughout the module as you learn about solving problems involving percent.

What Vocabulary Will You Learn?

Check the box next to each vocabulary term that you may already know.

☐ amount of error	☐ markdown	☐ principal
☐ commission	☐ markup	☐ sales tax
☐ discount	☐ percent error	☐ selling price
☐ fee	☐ percent of change	☐ simple interest
☐ gratuity	☐ percent of decrease	☐ tip
☐ interest	☐ percent of increase	☐ wholesale cost

Are You Ready?

Study the Quick Review to see if you are ready to start this module.
Then complete the Quick Check.

Quick Review

Example 1

Multiply with decimals.

Multiply $240 \times 0.03 \times 5$.

$240 \times 0.03 \times 5$

$= 7.2 \times 5$ — Multiply 240 by 0.03.

$= 36$ — Multiply 7.2 by 5.

Example 2

Write decimals as fractions and percents.

Write 0.35 as a fraction and as a percent.

$0.35 = 35\%$ — Move the decimal point two places to the right and write the percent sign.

$0.35 = \frac{35}{100}$ — There are 2 digits after the decimal point, so write those digits over 100.

$= \frac{7}{20}$ — Simplify.

Quick Check

1. Suppose Nicole saves $2.50 every day. How much money will she have in 4 weeks?

2. Approximately 0.92 of a watermelon is water. What percent and what fraction, in simplest form, represent this decimal?

How Did You Do?

Which exercises did you answer correctly in the Quick Check? Shade those exercise numbers at the right.

① ②

Percent of Change

I Can... use proportional relationships to solve percent of change problems.

Explore Percent of Change

Online Activity You will use bar diagrams to determine how a percent can be used to describe a change when a quantity increases or decreases.

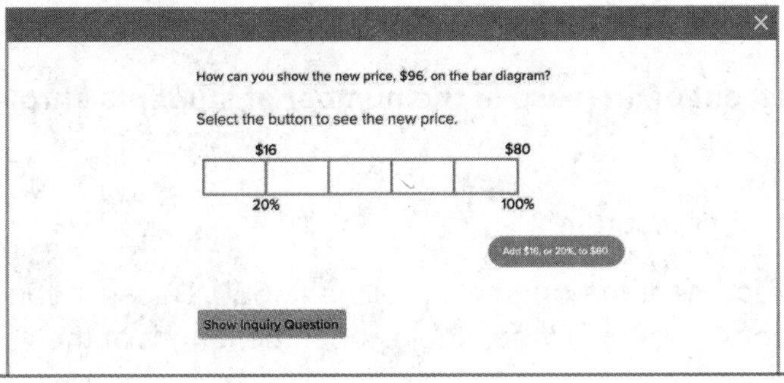

Learn Percent of Increase

A **percent of change** is a ratio, written as a percent, which compares the change in quantity to the original amount. If the original amount increased, then it is called a **percent of increase**.

The price of a monthly internet plan increased from $36 to $45. You can use a bar diagram to determine the percent of increase.

Draw a bar to represent the original price, $36. Because the original price is the whole, label the length of the bar 100%.

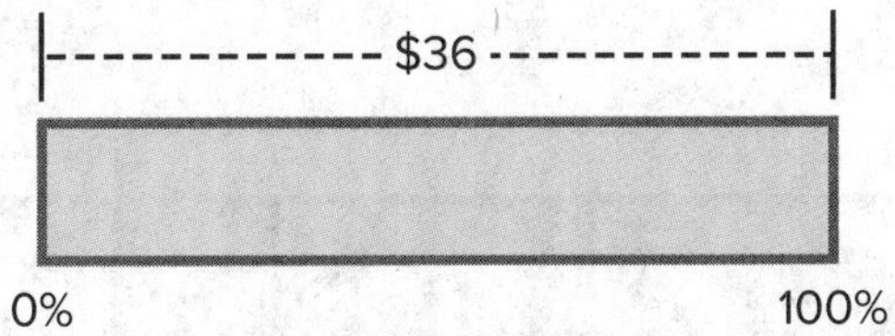

The price increased from $36 to $45, which is an increase of $9. Because $36 divided by $9 is 4, divide the bar representing $36 into four equal-size sections of $9 each. Each section represents 25% of the whole, $36.

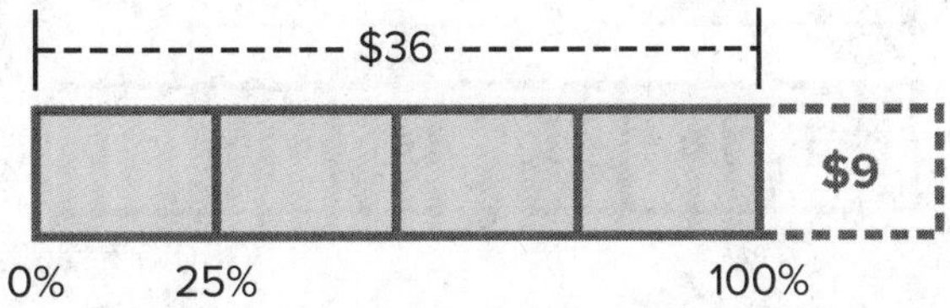

(continued on next page)

What Vocabulary Will You Learn?
percent of change
percent of decrease
percent of increase

Talk About It!
When finding percent of change, why is it important that the two quantities have the same unit of measure?

Your Notes

Each $9 section is 25% of the whole, $36. So, the price increased by 25%.

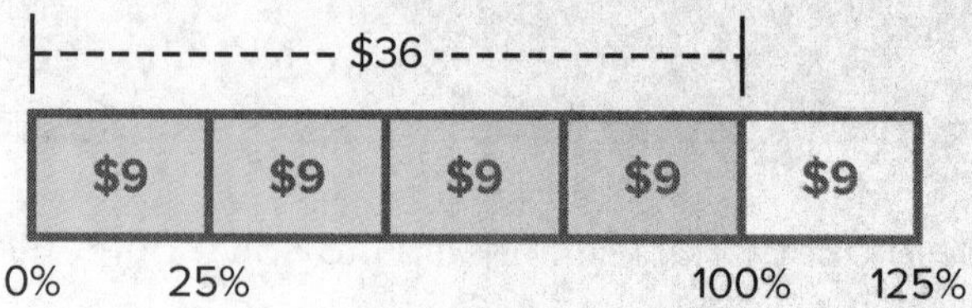

Note that the new price, $45, is 125% of the original price, but that the price increased by 25%.

Example 1 Percent of Increase

The enrollment at a middle school increased from 650 students to 780 students in five years.

What is the percent of increase in the number of students enrolled at the school?

Think About It!

By how many students did the enrollment increase?

Method 1 Use a bar diagram.

Draw a bar to represent the original enrollment, 650. Because the original number of students is the whole, label the length of the bar 100%.

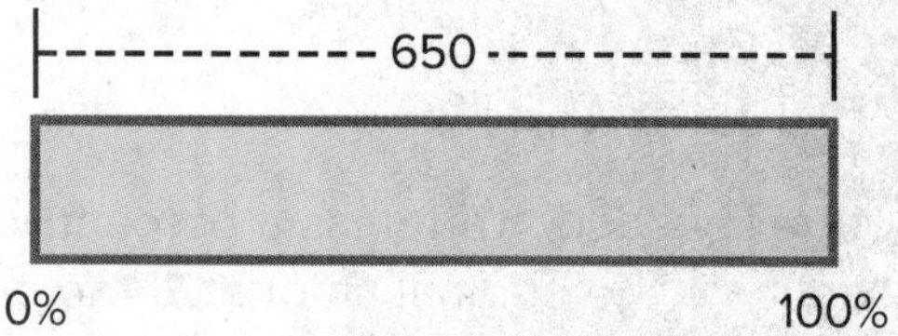

Enrollment increased from 650 to 780 students, which is an increase of 130 students. Divide the bar into equal parts so that the amount of increase, 130, is represented by one of the parts. Because 650 divided by 130 is 5, divide the bar representing 650 into five equal-size sections of 130 students each. Each section represents 20% of the whole, 650.

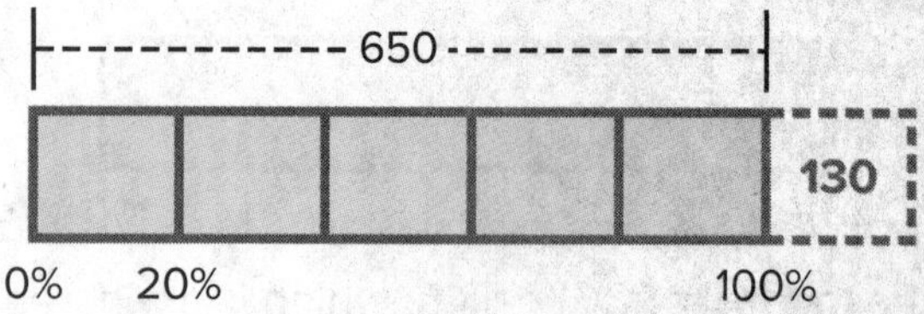

Each section that represents 130 students is 20% of the whole, 650. So, the enrollment increased by 20%.

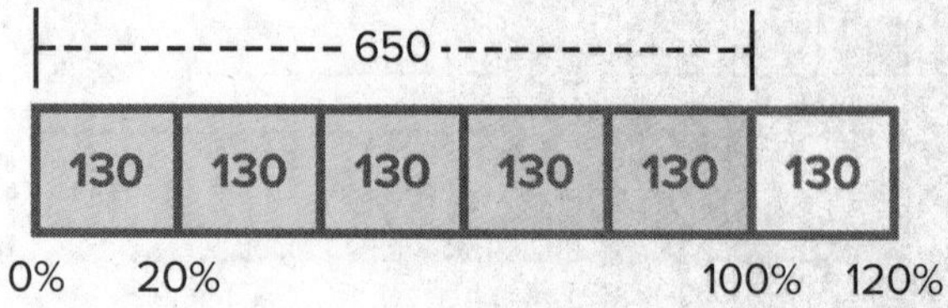

(continued on next page)

Method 2 Use equivalent ratios.

Step 1 Identify the part and the whole.

original amount = 650	This is the whole.
new amount = 780	This is the whole plus the part.
amount of increase = 130	This is the part.

Step 2 Find the percent of increase.

$\frac{\text{part}}{\text{whole}} = \frac{130}{650}$	Write the part-to-whole ratio. The part is 130. The whole is 650.
$= 0.20$	Divide.
$= \frac{20}{100}$	Write an equivalent ratio, as a rate per 100.
$= 20\%$	Definition of percent

So, using either method, the percent of increase in student enrollment is 20%.

Check

A certain brand of shampoo contains 12 ounces in a bottle. The company decides to redesign the shape of the shampoo bottle. The new type of bottle contains 15 ounces of shampoo. What is the percent of increase in the amount of shampoo the new bottle contains? Use any strategy.

Show your work here

Go Online You can complete an Extra Example online.

Pause and Reflect

Describe some examples of where you might see percent of change in your everyday life.

Record your observations here

Talk About It!

Why is the whole 650, and not 780?

Think About It!

Before you can find the percent of increase, what value do you need to calculate?

Talk About It!

Suppose the cost of a gallon of gas still increased by $1.80, but the original amount was $2.00 instead of $1.50. Would the percent of increase still be 120%? Explain.

Example 2 Percent of Increase

The average cost of gas in 2000 was about $1.50 per gallon. Suppose you purchase gas for $3.30 per gallon.

What is the percent of increase in the cost of a gallon of gas?

Step 1 Identify the part and the whole.

original amount = $1.50 — This is the whole.

new amount = $3.30 — This is the whole plus the part.

amount of increase = $1.80 — This is the part.

Step 2 Find the percent of increase.

$\frac{\text{part}}{\text{whole}} = \frac{1.80}{1.50}$ — Write the part-to-whole ratio. The part is 1.80. The whole is 1.50.

$= 1.20$ — Divide.

$= \frac{120}{100}$ — Write an equivalent ratio, as a rate per 100.

$= 120\%$ — Definition of percent

So, the percent of increase in the price per gallon of gas is 120%.

Check

Frankie calculated that she would need 9.5 yards of wallpaper to decorate her room. The wallpaper she chose had a large pattern, so instead she needed 13 yards of that paper. What is the percent of increase, rounded to the nearest percent, in the amount of wallpaper she needs?

Show your work here

Go Online You can complete an Extra Example online.

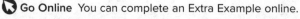

Learn Percent of Decrease

A percent of change is a ratio, written as a percent, which compares the change in quantity to the original amount. If the original amount is decreased, then the ratio is called a **percent of decrease.**

Jordan trained for a 5K race and his time decreased from 35 minutes to 28 minutes. You can use a bar diagram to determine the percent of decrease.

Draw a bar to represent the original time, 35 minutes. Because the original time is the whole, label the length of the bar 100%.

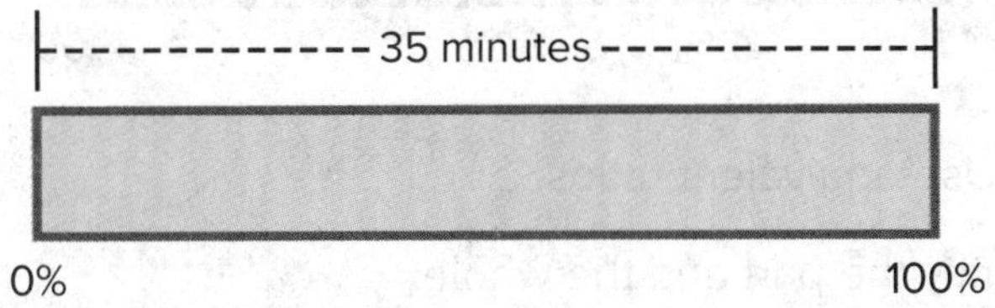

Jordan's time decreased from 35 minutes to 28 minutes, which is a decrease of 7 minutes. Because 35 divided by 7 is 5, divide the bar into five equal-size sections of 7 minutes each. Each section represents 20% of the whole, 35 minutes.

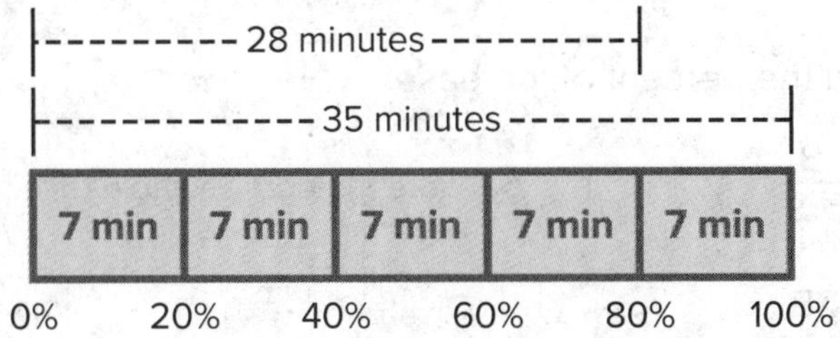

Each 7-minute section is 20% of the whole, 35. So, Jordan's time decreased by 20%.

Note that Jordan's new time is 80% of his original time, but that his time decreased by 20%.

Talk About It!

The percent of decrease is 20%. Explain why it is not 80%.

Example 3 Percent of Decrease

At the beginning of a chemistry experiment, the volume of liquid in a container was 25.2 milliliters. During the experiment, the volume dropped to 18.9 milliliters.

What is the percent of decrease in the volume of liquid?

Method 1 Use a bar diagram.

Draw a bar to represent the original volume, 25.2 milliliters. Because the original volume is the whole, label the length of the bar 100%.

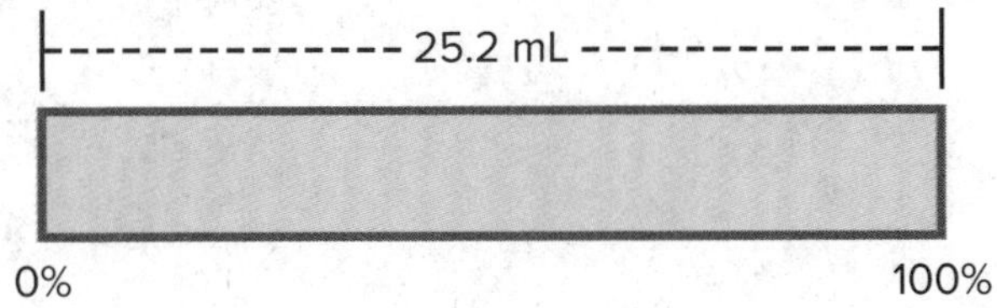

(continued on next page)

Think About It!

Before you can find the percent of decrease, what value do you need to calculate?

The volume decreased from 25.2 milliliters to 18.9 milliliters, which is a decrease of 6.3 milliliters. Because 25.2 divided by 6.3 is 4, divide the bar into four equal-size sections of 6.3 milliliters each. Each section represents 25% of the whole, 25.2 milliliters. So, the percent of decrease in the volume of the liquid is 25%.

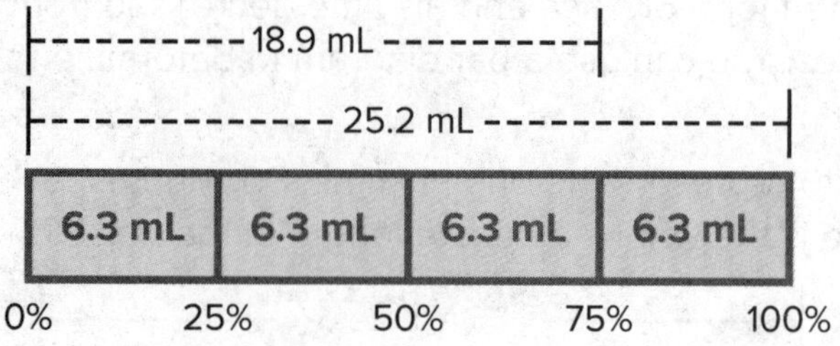

Method 2 Use equivalent ratios.

Step 1 Identify the part and the whole.

original amount = 25.2 — This is the whole.

new amount = 18.9 — This is the whole minus the part.

amount of decrease = 6.3 — This is the part.

Step 2 Find the percent of increase.

$\frac{\text{part}}{\text{whole}} = \frac{6.3}{25.2}$ — Write the part-to-whole ratio. The part is 6.3. The whole is 25.2.

$= 0.25$ — Divide.

$= \frac{25}{100}$ — Write an equivalent ratio, as a rate per 100.

$= 25\%$ — Definition of percent

So, the percent of decrease in the volume of the liquid is 25%.

Talk About It!

How can you use estimation to know that your answer is reasonable?

Check

When resting, a spring measures 73.3 millimeters. When the spring is compressed, it measures 51.29 millimeters. Find the percent of decrease in the length of the spring. Round to the nearest percent if necessary. Use any strategy.

Show your work here

Go Online You can complete an Extra Example online.

Apply Movies

The first known motion picture was filmed in 1888 and lasted for only 2.11 seconds. Today, we watch movies that last an average of about two hours. What is the percent of change in the times from 1888 to today? Round your answer to the nearest whole percent if necessary.

1 What is the task?

Make sure you understand exactly what question to answer or problem to solve. You may want to read the problem three times. Discuss these questions with a partner.

First Time Describe the context of the problem, in your own words.
Second Time What mathematics do you see in the problem?
Third Time What are you wondering about?

2 How can you approach the task? What strategies can you use?

Record your observations here

3 What is your solution?

Use your strategy to solve the problem.

Show your work here

4 How can you show your solution is reasonable?

Write About It! Write an argument that can be used to defend your solution.

Talk About It!
Why is the percent of change such a large number?

Check

The graph shows Alexa's bank account balance over the past four months. Between which consecutive months is the percent of increase the least?

Bank Account Balance

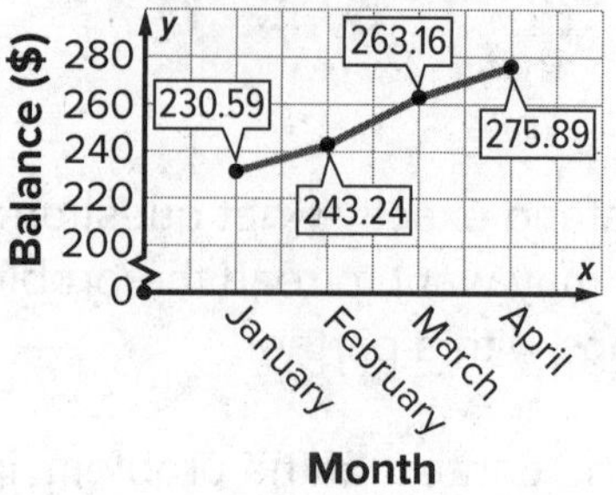

Ⓐ January and February

Ⓑ February and March

Ⓒ March and April

Show your work here

Go Online You can complete an Extra Example online.

Foldables It's time to update your Foldable, located in the Module Review, based on what you learned in this lesson. If you haven't already assembled your Foldable, you can find the instructions on page FL1.

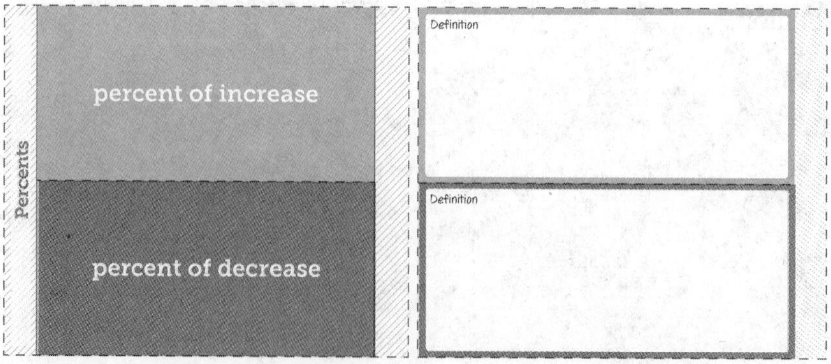

Name ______________________ Period __________ Date __________

Practice

Go Online You can complete your homework online.

Find each percent of change. Identify it as a percent of increase or decrease. **(Examples 1–3)**

1. 8 feet to 10 feet

2. 62 trees to 31 trees

3. 136 days to 85 days

4. Last month, the online price of a powered ride-on car was $250. This month, the online price is $330. What is the percent of increase for the price of the car? **(Example 1)**

5. At end of the first half of a football game, Nathan had carried the ball for 50.5 yards. By the end of the game, he carried the ball for a total of 75 yards. Find the percent of increase in the number of yards he carried. Round to the nearest whole tenth if necessary. **(Example 1)**

6. A music video website received 5,000 comments on a new song they released. The next day, the artist performed the song on television and an additional 1,500 comments were made on the website. What was the percent of increase? **(Example 1)**

7. When Ricardo was 9 years old, he was 56 inches tall. Ricardo is now 12 years old and he is 62 inches tall. Find the percent of increase in Ricardo's height to the nearest tenth. **(Example 1)**

8. At a garage sale, Petra priced her scooter for $15.50. She ended up selling it for $10.75. Find the percent of decrease in the price of the scooter. Round to the nearest tenth if necessary. **(Example 2)**

9. At the beginning of a baking session, there were 2.26 kilograms of flour in the bag. By the end of the baking session, there was 0.98 kilogram of flour in the bag. What is the percent of decrease, rounded to the nearest tenth, for the amount of flour? **(Example 2)**

Test Practice

10. Open Response The table shows the number of candid pictures of students for the yearbook for two consecutive years. What was the percent of decrease in the number of candid student pictures from 2015 to 2016, rounded to the nearest tenth?

Year	Number of Photos
2015	236
2016	214

Apply

11. The side length of the square shown is tripled. Which percent of increase is greater: the percent of increase for the perimeter of the square or the percent of increase for the area? How much greater?

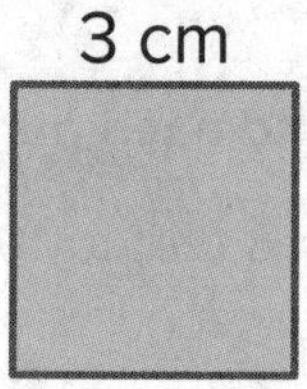

12. The Keatings have several bird feeders in their yard. They started with a 10.5-pound of birdseed. After 2 months, 12 ounces remained. What is the percent of change in the amount of birdseed after two months? Round your answer to the nearest tenth of a percent if necessary.

13. Create Write and solve a real-world problem involving a percent of increase with decimals.

14. Justify Conclusions Each of Mrs. White's two children received $50 for their savings account. The original amounts in their savings accounts were $500 and $300, respectively. Without calculating, which savings account had a greater percent of increase? Explain.

15. Reason Abstractly Determine if the following statement is *true* or *false*. Explain.

When the percent of change is a decrease, the original amount will be greater than the new amount.

16. Reason Abstractly Can a percent of change be greater than 100%? Explain.

Tax

I Can... use proportional relationships to find the amount of tax charged for an item.

Explore Sales Tax

Online Activity You will use Web Sketchpad to explore how sales tax changes the total cost to purchase an item.

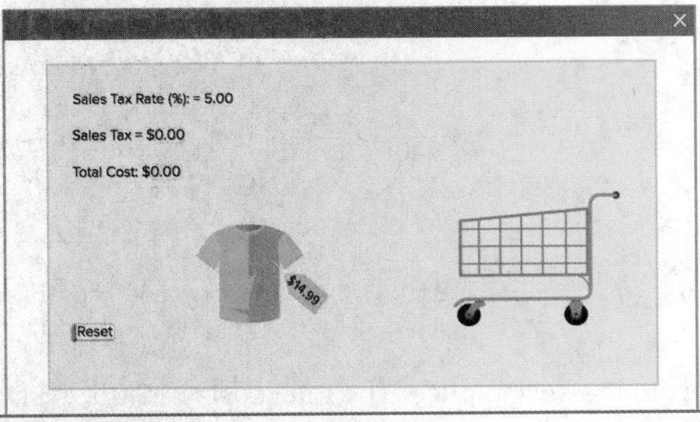

What Vocabulary Will You Learn?
sales tax

Learn Tax

A tax is an amount of money added to the price of certain goods and services. Tax is usually calculated as a percentage of the cost of the item. Some common forms of tax are sales tax, income tax, property tax, and hotel tax. Tax revenue is used to pay for government-provided services.

Sales tax is a state or local tax that is added to the price of an item or service. The total cost to purchase an item is the selling price plus the sales tax.

Sales tax varies depending on the city or state in which you live. Consider the cost of a $10 T-shirt with a sales tax rate of 7.5%. You can use equivalent ratios, and the percentage written as a rate per 100, to determine the amount of the tax. Let *t* represent the amount of tax.

sales tax on T-shirt → / cost of T-shirt →

$$\frac{t}{10} = \frac{7.5}{100} \quad \text{Percent}$$

$$\frac{0.75}{10} = \frac{7.5}{100} \quad (\div 10)$$

Because 100 ÷ 10 is 10, divide 7.5 by 10 to find *t*. The amount of tax is $0.75 or 75 cents.

Talk About It!
How can you find the total cost of the T-shirt, including the tax?

Your Notes

Think About It!
Before you can find the total Carie paid, what quantity do you need to find?

Talk About It!
Compare and contrast the two methods used for finding the total cost of the equipment.

Example 1 Sales Tax

Carie wants to buy sports equipment that costs \$140. The sales tax in her city is 5.75%.

What is the total cost of the equipment?

Method 1 Use ratio reasoning.

Write a proportion relating the two equivalent ratios. Let t represent the amount of sales tax. Then solve using ratio reasoning.

sales tax on sports equipment → / cost of sports equipment → $\frac{t}{140} = \frac{5.75}{100}$ } Percent

×1.40

$\frac{t}{140} = \frac{5.75}{100}$ Because $100 \times 1.40 = 140$, multiply 5.75 by 1.40 to find the value of t.

×1.40

$\frac{8.05}{140} = \frac{5.75}{100}$ $5.75 \times 1.40 = 8.05$, so, $t = 8.05$.

Add the sales tax to the selling price. The total cost is \$8.05 + \$140, or \$148.05.

Method 2 Use properties of operations. Let t represent the amount of sales tax.

$\frac{t}{140} = \frac{5.75}{100}$	Write the proportion.
$\frac{t}{140} = 0.0575$	Divide 5.75 by 100. A one-step equation results.
$140 \cdot \left(\frac{t}{140}\right) = (0.0575) \cdot 140$	Multiplication Property of Equality; Notice the tax is equal to the product of the percent, written as a decimal, and the cost.
$t = 8.05$	Simplify.

Add the sales tax to the selling price. The total cost is \$8.05 + \$140, or \$148.05.

So, using either method, the total cost of the sports equipment is \$148.05.

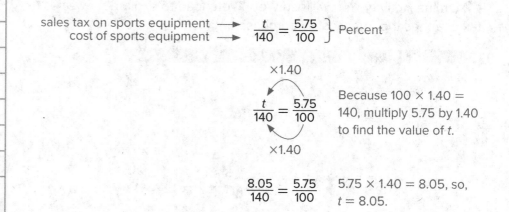

Check

Ashley is buying a laptop that sells for $749. The sales tax rate in her city is $8\frac{1}{4}$%. What is the total cost for the laptop? Round your answer to the nearest cent. Use any strategy.

Show your work here

Go Online You can complete an Extra Example online.

Think About It!

Is the tax less than, greater than, or equal to $28? How do you know?

Example 2 Hotel Tax

The cost of a hotel room rented for 2 nights is $280. There is also a 12% hotel room tax.

What is the total cost of the hotel room?

Method 1 Use ratio reasoning.

Write a proportion. Let *t* represent the amount of tax. Then solve using ratio reasoning.

tax on hotel room →
cost of hotel room → $\frac{t}{280} = \frac{12}{100}$ } Percent

$$\frac{t}{280} = \frac{12}{100} \quad (\times 2.80)$$

Because 100 × 2.80 = 280, multiply 12 by 2.80 to find the value of *t*.

$$\frac{33.60}{280} = \frac{12}{100}$$

12 × 2.80 = 33.60, so, *t* = 33.60.

Add the hotel tax to the cost of the hotel room. The total cost is $280 + $33.60, or $313.60.

(continued on next page)

Talk About It!

In Method 2, how could you use the steps in solving the equation to find the tax rate, or percentage, of any value?

Method 2 Use properties of operations. Let t represent the amount of tax.

$\frac{t}{280} = \frac{12}{100}$ Write the proportion.

$\frac{t}{280} = 0.12$ Divide 12 by 100. A one-step equation results.

$280 \cdot \left(\frac{t}{280}\right) = (0.12) \cdot 280$ Multiplication Property of Equality; Notice the tax is equal to the product of the percent, written as a decimal, and the cost.

$t = 33.60$ Simplify.

Add the hotel tax to the cost of the hotel room. The total cost is \$280 + \$33.60, or \$313.60.

So, using either method, the total cost of the hotel room is \$313.60.

Check

The cost of a hotel room for 5 nights is \$610. There is a 9.5% hotel tax. What is the total cost of the hotel room? Use any strategy.

Show your work here

Go Online You can complete an Extra Example online.

Pause and Reflect

Use the Internet, or another source, to research the sales tax rate for your city or state. Describe an item you may want to purchase and research the cost of the item. Trade with a partner and use the sales tax rate to find the total cost of the item, including sales tax.

Record your observations here

Example 3 Sales Tax

Henry purchases $56.00 worth of clothing at the store. The sales tax in his city is 7.5%.

What is the total cost of the clothing?

Method 1 Find the sales tax.

Use a proportion to find the amount of tax. Let *t* represent the amount of tax. Then add the tax to the cost of the clothing.

$\frac{t}{56} = \frac{7.5}{100}$ Write the proportion.

$\frac{t}{56} = 0.075$ Divide 7.5 by 100.

$56 \cdot \left(\frac{t}{56}\right) = (0.075) \cdot 56$ Multiplication Property of Equality; Notice the tax is equal to the product of the percent, written as a decimal, and the cost.

$t = 4.20$ Simplify.

Add the sales tax to the selling price.

$4.20 + $56.00 = $60.20 The total cost is $60.20.

Method 2 Add the sales tax percent to 100%.

Because 100% represents the selling price, add the sales tax percent to 100%. The total percent is 100% + 7.5% or 107.5%.

$\frac{c}{56} = \frac{107.5}{100}$ Write the proportion. Let *c* represent the total cost.

$\frac{c}{56} = 1.075$ Divide 107.5 by 100.

$56 \cdot \left(\frac{c}{56}\right) = (1.075) \cdot 56$ Multiplication Property of Equality; Notice the total cost is equal to the product of the total percent, written as a decimal, and the cost.

$c = 60.20$ Simplify.

So, the total cost of the clothing is $60.20.

(continued on next page)

Think About It!

How would you begin solving the problem?

Talk About It!

If you are at a store and need to quickly calculate the total with sales tax, what method would you use? Explain.

Method 3 Write an equation.

The total cost is the cost of the clothing plus the sales tax. Let c represent the total cost.

$c = 56 + 0.075(56)$	Write an equation. Write 7.5% as 0.075.
$= 56(1 + 0.075)$	Distributive Property
$= 56(1.075)$	Add.
$= 60.20$	Multiply.

Increasing the price by 7.5% is the same as multiplying the price by 1.075. So, using any method, the total cost of the clothing is $60.20.

Check

Jimena purchased $24 worth of crafting supplies. The tax rate in her city is 6.25%. What is the total cost of the crafting supplies? Use any strategy.

Show your work here

Go Online You can complete an Extra Example online.

Pause and Reflect

Compare what you have learned in this lesson to what you previously learned about proportional relationships.

Record your observations here

Apply Shopping

Davit goes shopping at his local grocery store and buys the items shown in the table. In the city he lives, there is a 7.3% sales tax on all items, except food. How much money does Davit spend at the grocery store? Round to the nearest cent if necessary.

Item	Cost ($)
Chicken	6.25
Carrots	1.99
Potatoes	2.55
Paper Plates	5.50
Napkins	6.15

1 What is the task?

Make sure you understand exactly what question to answer or problem to solve. You may want to read the problem three times. Discuss these questions with a partner.

First Time Describe the context of the problem, in your own words.
Second Time What mathematics do you see in the problem?
Third Time What are you wondering about?

2 How can you approach the task? What strategies can you use?

Record your observations here

3 What is your solution?

Use your strategy to solve the problem.

Show your work here

4 How can you show your solution is reasonable?

Write About It! Write an argument that can be used to defend your solution.

Talk About It!

How can you use estimation to determine if your answer is reasonable?

Check

The table shows the items that Victoria purchased at the market. In her city, there is an 8% sales tax on all items, except for food. How much money does Victoria spend at the market? Round to the nearest cent if necessary.

Item	Cost ($)
Cereal	3.55
Milk	3.10
Laundry Detergent	8.95
Yogurt	5.35
Dish Soap	2.95

Show your work here

Go Online You can complete an Extra Example online.

Pause and Reflect

How can you use mental math to estimate your total including sales tax?

Record your observations here

Name ______________________ Period __________ Date __________

Practice

Go Online You can complete your homework online.

Find the total cost to the nearest cent. **(Examples 1–3)**

1. \$18 breakfast; 7% tax

2. \$24 shirt; 6% tax

3. \$49.95 pair of shoes; 5% tax

4. Emily wants to buy new boots that cost \$68. The sales tax rate in her city is $5\frac{1}{2}$%. What is the total cost for the boots? **(Example 1)**

5. Jack wants to buy a coat that costs \$74.95. The sales tax rate in his city is $6\frac{1}{2}$%. What is the total cost for the coat? **(Example 1)**

6. Mr. Phuong stayed in a hotel room for 2 nights that cost \$210. The hotel room tax rate in the city is 12%. What is the total cost for the hotel room? **(Example 2)**

7. The cost of a hotel room during Lacy's trip is \$325. The hotel room tax in the city she is in is 10.5%. What is the total cost of the hotel room? **(Example 2)**

8. Robert spends \$30.45, before tax, at the bookstore. If the sales tax rate in his city is 7.25%, what is the total cost of his purchase? **(Example 3)**

Test Practice

9. **Multiple Choice** Anya purchased \$124.35 worth of home improvement items at the hardware store. If the sales tax rate in her city is 6.75%, what is the total cost of her purchase? **(Example 3)**

Ⓐ \$131.10

Ⓑ \$8.39

Ⓒ \$115.96

Ⓓ \$132.74

Apply

10. Jen purchased the items shown in the table. In the city she lives in, the sales tax rate is 7.15%. In another city, the sales tax rate is 6.35%. How much more is she spending if she purchases the items in the city she lives in? Round to the nearest cent.

Item	Cost ($)
shirt	11.99
shoes	35.50
belt	6.75
socks	3.00

11. Shamir is trying to decide between two cities to travel to for a weekend trip. The prices for a hotel room for two nights and the hotel tax rate are listed in the table. What is the difference in cost between the two cities for a weekend trip? Round to the nearest cent.

City	Cost ($)	Tax Rate
A	250	12.5%
B	215	14.75%

12. MP **Find the Error** A student is finding the total cost c, including sales tax, of a book that costs \$9.95. The sales tax rate is 9%. Find the student's mistake and correct it.

$c = 1.9 \times 9.95$

$c \approx 18.91$

$c \approx \$18.91$

13. MP **Identify Structure** Write two different expressions to find the total cost of an item that costs \$$a$ if the sales tax is 6%. Explain why the expressions give the same result.

14. What are some similarities between sales tax and hotel tax?

15. MP **Reasoning** A dollhouse is on sale for \$160. The tax rate is $5\frac{1}{4}$%. Without calculating, will the tax be greater than or less than \$8? Write an argument that can be used to defend your solution.

Tips and Markups

I Can... use proportional relationships to find the amount to pay for a tip and the amount of markup on items.

What Vocabulary Will You Learn?
gratuity
markup
selling price
tip
wholesale cost

Learn Tips

A **tip,** or **gratuity,** is an additional amount of money paid in return for a service. This amount is sometimes a percent of the service cost. The total amount paid is the cost of the service plus the tip.

Example 1 Tips

The bill for a group of eight people at a restaurant was $125 before the tip was added. The group wants to add an 18% tip.

What will be the total bill including the tip?

Method 1 Use ratio reasoning.

Write a proportion. Then solve using ratio reasoning. Let t represent the amount of the tip.

amount of tip → / amount of bill → $\frac{t}{125} = \frac{18}{100}$ } Percent

×1.25

$\frac{t}{125} = \frac{18}{100}$ Because $100 \times 1.25 = 125$, multiply 18 by 1.25 to find the value of t.

×1.25

$\frac{22.50}{125} = \frac{18}{100}$ $18 \times 1.25 = 22.50$, so, $t = 22.50$.

Add the tip to the bill. The total cost is $125 + $22.50, or $147.50.

Method 2 Use properties of operations. Let t represent the amount of the tip.

$\frac{t}{125} = \frac{18}{100}$ Write the proportion.

$\frac{t}{125} = 0.18$ Divide 18 by 100. A one-step equation results.

$125 \cdot \left(\frac{t}{125}\right) = (0.18) \cdot 125$ Multiplication Property of Equality

$t = 33.60$ Simplify.

Add the tip to the bill. The total cost is $125 + $22.50, or $147.50. So, using either method, the total cost of the bill is $147.50.

Think About It!
What is a good estimate for the solution? Explain how you calculated the solution.

Talk About It!
How could you solve the problem another way?

Your Notes

Check

Amy wants to tip her hairstylist 20% for a haircut that costs \$48. What is her total bill with tip? Use any strategy.

Show your work here

Go Online You can complete an Extra Example online.

Learn Markup

In order to make a profit, stores typically sell items for more than what they pay for them. The amount the store pays for an item is called the **wholesale cost**. The amount of increase is called the **markup**. The **selling price** is the amount the customer pays for an item. The selling price is equal to the wholesale cost plus the markup.

selling price = wholesale cost + markup

Think About It!
What is a good estimate for the solution? Explain how you calculated that estimate.

Example 2 Markup

The wholesale cost for each shirt at a clothing store is \$17. The store manager plans to mark up the shirts by 125%.

What will be the selling price for each shirt?

Method 1 Use ratio reasoning.

Write a proportion. Then solve using ratio reasoning. Let x represent the selling price.

amount of markup → x, wholesale cost → 17: $\frac{x}{17} = \frac{125}{100}$ } Percent

$$\frac{x}{17} = \frac{125}{100} \quad (\times 0.17)$$

Because $100 \times 0.17 = 17$, multiply 125 by 0.17 by to find the value of x.

$$\frac{21.25}{17} = \frac{125}{100}$$

$125 \times 0.17 = 21.25$, so, $x = 21.25$.

Add the markup to the wholesale cost. The selling price is \$17 + \$21.25, or \$38.25.

(continued on next page)

Method 2 Use properties of operations.

$\frac{x}{17} = \frac{125}{100}$ Write the proportion. Let x represent the selling price.

$\frac{x}{17} = 1.25$ Divide 125 by 100. A one-step equation results.

$17 \cdot \left(\frac{x}{17}\right) = (1.25) \cdot 17$ Multiplication Property of Equality

$x = 21.25$ Simplify.

Add the markup to the wholesale cost. The selling price is $17 + $21.25, or $38.25.

So, using either method, the selling price is $38.25.

Check

The wholesale cost for a basketball backboard is $32. If the markup is $85\frac{1}{2}$%, what is the selling price? Use any strategy.

Show your work here

Go Online You can complete an Extra Example online.

Example 3 Markup

Ben's family is shopping for a new car. The selling price of a car is $24,199.50. Ben researches to find that the wholesale cost of the car is $22,100.00.

What is the percent of markup?

Finding the percent of markup is the same as finding the percent of increase.

Step 1 Identify the part and the whole.

original amount = $22,100.00 This is the whole.

new amount = $24,199.50 This is the whole plus the part.

amount of increase = $2,099.50 This is the part.

Talk About It!

Compare the wholesale cost with the selling price. How do you know the selling price is reasonable?

Think About It!

What is a good estimate for the solution? Explain how you calculated that estimate.

Step 2 Find the percent of increase.

$\frac{\text{part}}{\text{whole}} = \frac{2{,}099.50}{22{,}100.00}$ Write the part-to-whole ratio. The part is 2,099.50. The whole is 22,100.00.

$= 0.095$ Divide.

$= \frac{9.5}{100}$ Write an equivalent ratio, as a rate per 100.

$= 9.5\%$ Definition of percent

So, the percent of markup for the wholesale price of the car is 9.5%.

Talk About It!

How is finding the percent of markup different than finding the selling price of an item?

Check

Mikka is making jewelry for a craft show. The wholesale cost of a bracelet is $12.50. If she sells them for $20, what is the percent of markup?

Show your work here

Go Online You can complete an Extra Example online.

Pause and Reflect

Compare and contrast tips and markups. Where have you seen or used tips and markups in your everday life?

Record your observations here

Apply Dining Out

Kerry and three friends went out for dinner. They split a large pizza, and each person had a salad and a soda. They want to leave a 15% tip on the cost of the food, and the sales tax is $8\frac{1}{4}$%. How much will each person pay if they split the bill evenly?

Pizza	$18.60
Salad	$2.50
Soda	$2.25

1 What is the task?

Make sure you understand exactly what question to answer or problem to solve. You may want to read the problem three times. Discuss these questions with a partner.

First Time Describe the context of the problem, in your own words.
Second Time What mathematics do you see in the problem?
Third Time What are you wondering about?

2 How can you approach the task? What strategies can you use?

Record your observations here

3 What is your solution?

Use your strategy to solve the problem.

Show your work here

Talk About It!
What steps should you take before splitting the bill?

4 How can you show your solution is reasonable?

Write About It! Write an argument that can be used to defend your solution.

Check

Brian has $24 worth of pizza delivered to his house. He pays the bill plus a 15% tip and 7% sales tax. He also pays a $3 delivery fee that is charged after the tax and tip. How much change does he receive, if he pays with two $20 bills?

Show your work here

Go Online You can complete an Extra Example online.

Pause and Reflect

Explain how tips and markups are percents of increase.

Record your observations here

Name ____________________ Period __________ Date __________

Practice

Go Online You can complete your homework online.

Find the total cost to the nearest cent. Use any strategy. (Examples 1 and 2)

1. \$20 haircut; 10% tip

2. \$24 lunch; 15% tip

3. \$185 TV; 5% markup

4. Vera went to the local salon to get a haircut. The cost was \$24. Vera tipped the hair stylist 18%. What was the total cost of haircut including the tip? Round to the nearest cent. **(Example 1)**

5. The Gomez family ordered \$39.50 worth of pizza and subs. They gave the delivery person a 20% tip. What was the total cost of the food and tip? Round to the nearest cent. **(Example 1)**

6. The wholesale cost of a bicycle is \$98.75. The markup for the bicycle is 33.3%. Find the selling price of the bicycle. Round to the nearest cent. **(Example 2)**

7. The wholesale cost for a purse in a department store is \$12.50. The store plans to mark up the purse by 140%. What will be the selling price of the purse? Round to the nearest cent. **(Example 2)**

8. Keri is making doll clothes for a holiday craft show. The wholesale cost of the materials for one outfit is \$9.38. If she sells an outfit for \$15, what is the percent of markup? Round to the nearest percent. **(Example 3)**

9. A pet store sells a large dog kennel for \$98.50. The wholesale cost of the kennel is \$63.55. What is the percent of markup? Round to the nearest percent. **(Example 3)**

Test Practice

10. Open Response An elementary school wants to purchase a new swing set. The table shows the selling price of the swing sets they are interested in buying. The markup for both swing sets is $20\frac{1}{4}$%. The school decides to buy the Adventurers swing set. What is the selling price of the swing set they are buying?

Swing Set	Wholesale Price (\$)
Adventurers	3,056
Thunder Ridge	4,125

Apply

11. Louise bought a frame and canvas online for $125. Then she bought the supplies shown in the table and paid a 7% sales tax on these supplies only. She took the amount that she spent on the canvas, frame, and supplies, and marked up the price by 115%. How much did she charge for her painting? Round to the nearest cent.

Item	Cost ($)
Brushes	12.50
Paint	19.50

12. Brendan has $65 worth of balloons and flowers delivered to his mother. He pays the bill plus an 8.5% sales tax and an 18% tip on the total cost including tax. He also pays a $10 delivery fee that is charged after the tax and tip. How much change does he receive if he pays with two $50 bills? Round to the nearest cent.

13. Toru takes his dog to be groomed. The fee to groom the dog is $70 plus a 10% tip. Is $80 enough to pay for the service and tip? Explain your reasoning.

14. MP **Use a Counterexample** Determine if the following is statement is *true* or *false*. If false, provide a counterexample.

It is impossible to increase the cost of an item by less than 1%.

15. MP **Identify Structure** Write two different expressions to find the total cost of a service that costs $*x* if the tip is 18%. Explain why the expressions give the same result.

16. **Create** Write and solve a real-world problem in which you find sales tax and a tip.

Discounts

I Can... use proportional relationships to find the amount of discount or markdown.

What Vocabulary Will You Learn?
discount
markdown

Learn Discounts

Discount or **markdown** is the amount by which the regular price of an item is reduced. The sale price is the original price minus the discount. The percent of discount is a percent of decrease.

sale price = original price − discount

Example 1 Discounts

Summer Sports is having a sale. A volleyball has an original price of \$59. It is on sale for 65% off the original price.

What is the sale price of the volleyball?

Method 1 Find the amount of the discount. Let d represent the amount of the discount.

Write a proportion. Then solve using ratio reasoning.

amount of discount → / original price → $\frac{d}{59} = \frac{65}{100}$ } Percent

$\frac{d}{59} \underset{\times 0.59}{\overset{\times 0.59}{=}} \frac{65}{100}$ Because $100 \times 0.59 = 59$, multiply 65 by 0.59 to find the value of d.

$\frac{38.35}{59} = \frac{65}{100}$ $65 \times 0.59 = 38.35$, so, $d = 38.35$.

Subtract the discount from the original price. The sale price is \$59 − \$38.35, or \$20.65.

Talk About It!
In Method 1, why is the answer not \$38.35?

(continued on next page)

Your Notes

Method 2 Subtract the discount percent from 100%.

Because 100% represents the original price, subtract the percent of the discount from 100%. The sale price is 100% − 65% or 35% of the original price. Then write a proportion and solve using ratio reasoning. Let x represent the sale price.

$$\frac{x}{59} = \frac{35}{100}$$

sale price → x; original price → 59; $\frac{35}{100}$ → Percent

$$\frac{x}{59} \overset{\times 0.59}{=} \frac{35}{100}$$

Because $100 \times 0.59 = 59$, multiply 35 by 0.59 to find the value of x.

$$\frac{20.65}{59} = \frac{35}{100}$$

$35 \times 0.59 = 20.65$, so, $x = 20.65$.

So, using either method, the sale price is $20.65.

Check

A restaurant decreased their prices for a day to their prices from 1964. A pizza that usually sells for $15.40 was marked down 85%. What was the price of the pizza in 1964? Use any strategy.

Show your work here

Go Online You can complete an Extra Example online.

Think About It!

How can you estimate the solution?

Example 2 Combined Discounts

During a clearance sale at an electronics store, certain tablets were marked down 20%. One day, an additional 30% was taken off already-reduced prices. A tablet originally sold for $375.

What was the final price after both discounts were applied?

Step 1 Find the price after the first discount.

Because 100% represents the original price, subtract the percent of the original discount from 100%. The sale price is 100% − 20% or 80% of the original price. Then write a proportion and solve using ratio reasoning. Let x represent the sale price after the first discount.

(continued on next page)

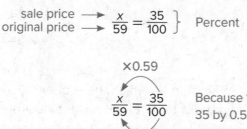

$$\text{sale price} \longrightarrow \frac{x}{375} = \frac{80}{100} \Bigg\} \quad \text{Percent}$$
$$\text{original price} \longrightarrow$$

$$\overset{\times 3.75}{\curvearrowright}$$
$$\frac{x}{375} = \frac{80}{100} \qquad \text{Because } 100 \times 3.75 = 375, \text{ multiply}$$
$$\underset{\times 3.75}{\curvearrowleft} \qquad\qquad 80 \text{ by } 3.75 \text{ to find the value of } x.$$

$$\frac{300}{375} = \frac{80}{100} \qquad 80 \times 3.75 = 300, \text{ so, } x = 300. \text{ The}$$
$$\text{price after the first discount is } \$300.$$

Step 2 Find the final price after the additional discount.

Because 100% now represents the price after the first discount, subtract the percent of the additional discount from 100%. The sale price is 100% − 30% or 70% of the price after the first discount. Then write a proportion and solve using ratio reasoning. Let x represent the sale price after the additional discount.

$$\text{sale price} \longrightarrow \frac{x}{300} = \frac{70}{100} \Bigg\} \quad \text{Percent}$$
$$\text{original price} \longrightarrow$$

$$\overset{\times 3}{\curvearrowright}$$
$$\frac{x}{300} = \frac{70}{100} \qquad \text{Because } 100 \times 3 = 300, \text{ multiply}$$
$$\underset{\times 3}{\curvearrowleft} \qquad\qquad 70 \text{ by } 3 \text{ to find the value of } x.$$

$$\frac{210}{300} = \frac{70}{100} \qquad 70 \times 3 = 210, \text{ so, } x = 210.$$

So, the final price of the tablet after both discounts were applied was $210.

Check

A clothing store marked all of their summer clothes down 50%. A sign in the store indicates that an additional 20% is to be taken off clearance prices. What is the final price of a top that originally sold for $58?

Show your work here

🅑 **Go Online** You can complete an Extra Example online.

Talk About It!

The final price had a discount of 20% followed by a discount of 30%. Is this the same as finding 20% + 30% or 50% of the original price? Use the values in the Example to justify your reasoning.

Think About It!

Are you trying to find the part, percent, or whole?

Talk About It!

When writing the proportion, why is the percent 75 out of 100 instead of 25 out of 100?

Example 3 Find the Original Price

Sandy has a 25% discount coupon for athletic equipment. She buys hockey equipment for a final price of \$172.50.

What is the original price?

If the amount of the discount is 25%, the sale price is 100% − 25%, or 75%, of the original price. So, 75% of the original price is \$172.50.

Write a proportion and solve using ratio reasoning. Let x represent the original price.

sale price → / original price →

$$\frac{172.50}{x} = \frac{75}{100} \text{ Percent}$$

$$\frac{172.50}{x} = \frac{75}{100} \quad (\times 2.3)$$

Because $75 \times 2.3 = 172.50$, multiply 100 by 2.3 to find the value of x.

$$\frac{172.50}{230} = \frac{75}{100}$$

$100 \times 2.3 = 230$, so, $x = 230$.

So, the original price of the hockey equipment is \$230.

Check

A pair of running shoes is on sale for \$125.99. If the sale price is discounted 9% from the original price, what is the original price? Round to the nearest cent.

Show your work here

Go Online You can complete an Extra Example online.

Apply Shopping

The Wares want to buy a new computer. Store A has a regular price of $1,300 and is offering a discount of 20%. Store B has a regular price of $1,089 with no discount. They want to purchase the computer that is less expensive. If there is a $7\frac{1}{4}$% sales tax in their city, at which store should they buy their computer, and how much money will they save if they buy at that store instead of the other store?

Go Online watch the animation.

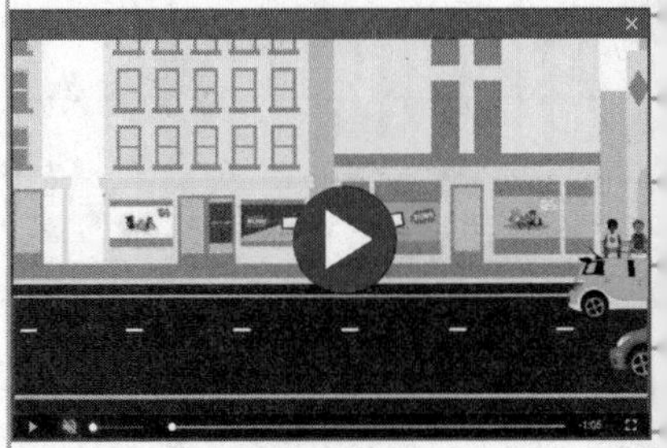

1 What is the task?

Make sure you understand exactly what question to answer or problem to solve. You may want to read the problem three times. Discuss these questions with a partner.

First Time Describe the context of the problem, in your own words.
Second Time What mathematics do you see in the problem?
Third Time What are you wondering about?

2 How can you approach the task? What strategies can you use?

Record your observations here

3 What is your solution?

Use your strategy to solve the problem.

Show your work here

4 How can you show your solution is reasonable?

Write About It! Write an argument that can be used to defend your solution.

Talk About It!
How can you quickly determine which computer will be cheaper before the sales tax is applied?

Check

A bike shop has two road bikes on sale. The Road Warrior has a regular price of $379 and is discounted 25%. The Road Racer has a regular price of $320 and is discounted 12%. If the sales tax rate is 6.5%, which bike has the less expensive sale price? How much will you save by buying that bike?

Show your work here

Go Online You can complete an Extra Example online.

Pause and Reflect

Write a paragraph explaining why a 30% discount is not the same as a 20% discount plus an additional 10% discount. Which is the better discount?

Record your observations here

Name ______________________ Period ____________ Date ____________

Practice

Go Online You can complete your homework online.

Find the sale price to the nearest cent. Use any strategy. **(Example 1)**

1. $140 coat; 10% discount

2. $80 boots; 25% discount

3. $325 tent; 15% discount

4. A toy store is having a sale. A video game system has an original price of $99. It is on sale for 40% off the original price. Find the sale price of the game system. Round to the nearest cent. **(Example 1)**

5. A yearly coffee club subscription costs $65. Avery received an offer for 62% off the subscription cost. What is the sale price of the subscription? Round to the nearest cent. **(Example 1)**

6. During a clearance sale at a sporting goods store, skateboards were marked down 30%. On Saturday, an additional 25% was taken off already reduced prices of skateboards. If a skateboard originally cost $119.50, what was the final price after all discounts had been taken? Round to the nearest cent. **(Example 2)**

7. At an electronics store, a smart phone is on sale for 35% off the original price of $679. If you use the store credit card, you can receive an additional 15% off the sale price. What is the final price of the smart phone if you use the store credit card? Round to the nearest cent. **(Example 2)**

8. Gary had a 40% discount for new tires. The sale price of a tire was $96.25. What was the original price of the tire? Round to the nearest cent. **(Example 3)**

9. A swimsuit is on sale for $45.50. If the sale price is discounted 5% from the original price, what was the original price? Round to the nearest cent. **(Example 3)**

Test Practice

10. Open Response A shoe store is having a clearance sale on their summer shoes. All summer shoes are marked 55% off. A sign states you can take an additional 10% off the clearance sale prices. Kelly is deciding between two pairs of sandals shown in the table. If she buys the blue sandals, what is the final price Kelly will pay? Round to the nearest cent.

Shoes	Original Price
Blue Sandals	$75
Tan Sandals	$68

Apply

11. A recreational outlet has two trampolines on sale. The table shows the original prices. The Skye Bouncer is discounted 15% and the Ultimate is discounted 13%. If the sales tax rate is 7.5%, which trampoline has the lower sale price? How much will you save by buying that trampoline? Round to the nearest cent.

Trampoline Model	Original Price ($)
Skye Bouncer	1,480
Ultimate	1,450

12. Pets Plus and Pet Planet are having a sale on the same aquarium. At Pets Plus the aquarium is on sale for 30% off the original price and at Pet Planet it is discounted by 25%. If the sales tax rate is 8%, which store has the lower sale price? How much will you save by buying the aquarium there? Round to the nearest cent.

Store	Original Price of Aquarium ($)
Pets Plus	118
Pet Planet	110

13. MP **Persevere with Problems** Suppose an online store has an item on sale for 10% off the original price. By what percent does the store have to increase the price of the item in order to sell it for the original amount? Explain.

14. MP **Identify Structure** A shoe store buys packs of socks wholesale for $5 each and marks them up by 40%. The store decides to discount the packs of socks by 40%. Is the discounted price $5? Explain your reasoning.

15. A model car is for sale online. The owner discounts the price by 5% each day until it sells. On the third day the car sells. If the original price of the model car was $40, how much did the car sell for? What percent discount does this represent from the original price?

16. Describe a real-world problem where an item is discounted by 75%. Then find the original price.

Interest

I Can... use the simple interest formula to find the amount of interest earned for a given principal, at a given interest rate, for a given period of time.

What Vocabulary Will You Learn?
interest
principal
simple interest

Explore Interest

Online Activity You will use eTools to explore the simple interest formula.

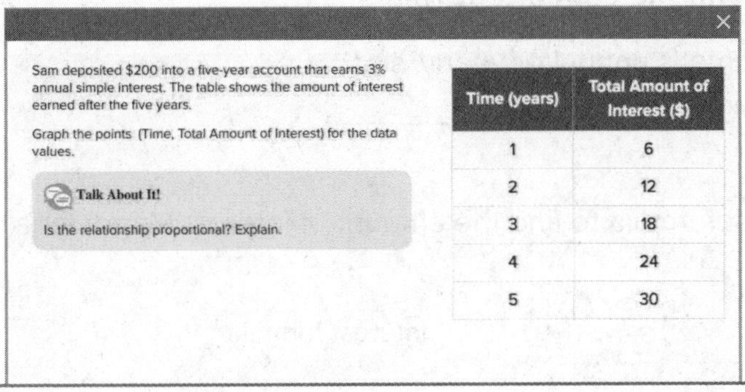

Sam deposited $200 into a five-year account that earns 3% annual simple interest. The table shows the amount of interest earned after the five years.
Graph the points (Time, Total Amount of Interest) for the data values.
Talk About It!
Is the relationship proportional? Explain.
Time (years)
Total Amount of Interest ($)
1 6
2 12
3 18
4 24
5 30

Learn Simple Interest

If you borrow money or deposit money, the **principal** is the amount of money borrowed or deposited. **Interest** is the amount paid or earned for the use of the principal. **Simple interest** is calculated using specified periods of time.

Percents are used to calculate interest. You can earn interest by letting the bank use the money you deposit in a bank account. You will pay interest if you borrow money from the bank or use a credit card.

Go Online Watch the animation to find the interest earned for the information in the table.

principal	$430
interest rate	2% or 0.02
time	5 years

The animation shows that the steps for finding the interest are:

$I = prt$	Interest formula
$I = 430 \cdot 0.02 \cdot 5$	principal is $430, rate is 0.02, and time is 5 years
$I = 43$	Simplify.

So, the interest is $43. This means that, in 5 years, you will have $430 + $43, or $473, in your bank account if you do not deposit or withdraw in those 5 years.

(continued on next page)

Your Notes

Words	Symbols
Simple Interest I is the product of the principal p, interest rate r, and the length of time t, expressed in years.	$I = prt$

Example 1 Find Simple Interest

Magdalena put $580 into a savings account. The account pays 2.5% simple interest each year.

If she neither adds nor withdraws money from the account, how much interest will she earn after 2 years?

The period of time is annual or yearly, so $t =$ ______.

The rate is 2.5% or, as a decimal, ______.

Use the Interest formula to find the amount of interest Magdalena will earn.

$I = prt$	Interest formula
$I = \square \cdot \square \cdot \square$	The principal is 580, the rate is 0.025, and the time is 2.
$I = 29$	Multiply.

So, Magdalena earned $ ______ in interest in 2 years.

Check

Curtis deposits $550 into a savings account. The account pays 1.75% simple interest on an annual basis. If he does not add or withdraw money from the account, how much interest will he earn after 3 years?

Show your work here

Go Online You can complete an Extra Example online.

Example 2 Find Simple Interest

Suppose Magdalena opens a savings account for only 6 months. She puts $580 in that account and earns 3.5% each year.

How much interest will she earn?

Because the period of time is annual or yearly, 6 months equals $\frac{6}{12}$ or

_________ year.

What is the rate 3.5% written as a decimal? _________

Use the Interest formula to find the amount of interest Magdalena will earn.

$I = prt$	Interest formula
$I = \square \cdot \square \cdot \square$	The principal is 580, the rate is 0.035, and the time is 0.5.
$I = 10.15$	Multiply.

So, Magdalena earned $ _________ in interest in _________ months.

Check

Carrie invests $430 into a savings account. The account pays 2.5% simple interest on an annual basis. If she does not add or withdraw money from the account, how much interest will she earn after 15 months?

Show your work here

Go Online You can complete an Extra Example online.

Think About It!

What is the simple interest formula? Are you trying to find the *interest, principal, rate,* or *time*?

Talk About It!

How is the process of finding simple interest different when the time is given in months?

Example 3 Find Simple Interest

Think About It!

How would you begin solving the problem?

Rondell's parents are starting a real estate business. They borrow $99,400 from the bank to finance their start-up operations. The simple interest rate is $4\frac{3}{4}\%$ per year, and they plan to take 15 years to repay the loan.

How much simple interest will they pay?

The period of time is annual or yearly, so $t =$ ________.

What is the rate $4\frac{3}{4}\%$ written as a decimal? ________

Use the Interest formula to find the amount of interest they will pay.

$I = prt$	Interest formula
$I = \square \cdot \square \cdot \square$	The principal is 99,400, the rate is 0.0475, and the time is 15.
$I = 70{,}822.50$	Multiply.

So, Rondell's parents will pay $ ____________ in interest on the loan.

Check

Abbie borrows $3,216 at a rate of $6\frac{4}{5}\%$ per year. How much simple interest will she owe if it takes 2 years to repay the loan?

Show your work here

Go Online You can complete an Extra Example online.

Pause and Reflect

Why is it important to pay off loans as soon as possible?

Record your observations here

Apply Car Shopping

Alex is buying a car that costs $18,000. The tax rate is 7%, and he plans to make a down payment of $2,000. The sales tax is added to the price of the car before the down payment is made. He is considering a five-year loan with a simple interest rate of 6.2% each year. What will be his monthly payment?

1 What is the task?

Make sure you understand exactly what question to answer or problem to solve. You may want to read the problem three times. Discuss these questions with a partner.

First Time Describe the context of the problem, in your own words.
Second Time What mathematics do you see in the problem?
Third Time What are you wondering about?

2 How can you approach the task? What strategies can you use?

Record your observations here

Talk About It!

How can the amount of a down payment affect the monthly payment?

3 What is your solution?

Use your strategy to solve the problem.

Show your work here

4 How can you show your solution is reasonable?

Write About It! Write an argument that can be used to defend your solution.

Check

Donte is buying a car that costs $8,000. He is deciding between a 3-year loan and a 4-year loan. Use the rates in the table to determine how much money he will save if he chooses the 3-year loan instead of the 4-year loan.

Time (y)	Simple Interest Rate (%)
3	2.25
4	2.5
5	3

Show your work here

Go Online You can complete an Extra Example online.

Pause and Reflect

Have you ever wondered when you might use the concepts you learn in math class? What are some everyday scenarios in which you might use what you learned today?

Record your observations here

Name ______________________ Period ____________ Date ____________

Practice

Go Online You can complete your homework online.

Find the simple interest earned, to the nearest cent, for each principal, interest rate, and time. **(Example 1)**

1. \$530, 6%, 1 year

2. \$1,200, 3.5%, 2 years

3. \$750, 7%, 3 years

4. Elena's father put \$460 into a savings account for her. The account pays 2.5% simple interest each year. If he neither adds nor withdraws money from the account, how much interest will the account earn after 4 years? Round to the nearest cent. **(Example 1)**

5. Ethan put \$1,250 into a savings account. The account pays 4.5% simple interest on an annual basis. If he does not add or withdraw money from the account, how much interest will he earn after 2 years? Round to the nearest cent. **(Example 1)**

6. Marc deposits \$840 into a savings account. The account pays 2% simple interest on an annual basis. If he does not add or withdraw money from the account, how much interest will he earn after 6 months? Round to the nearest cent. **(Example 2)**

7. Nina's grandmother deposits \$3,000 into a savings account for her. The account pays 5.5% simple interest on an annual basis. If she does not add or withdraw money from the account, how much interest will she earn after 21 months? Round to the nearest cent. **(Example 2)**

8. Jack borrows \$2,700 at a rate of 8.2% per year. How much simple interest will he owe if it takes 3 months to repay the loan? Round to the nearest cent. **(Example 3)**

9. Liliya's parents borrow \$1,400 from the bank for a new washer and dryer. The interest rate is 7.5% per year. How much simple interest will they pay if they take 18 months to repay the loan? Round to the nearest cent. **(Example 3)**

Test Practice

10. Open Response The table shows the interest rates for auto repair loans based on how long it takes to pay off the loan. Jin borrows \$3,600 and plans to pay the loan off in 18 months. How much simple interest will he owe if it takes 18 months to repay the loan? Round to the nearest cent.

Time	Rate (%)
6 months	3.5
12 months	4.0
18 months	4.25

Apply

11. Jarvis is buying a boat that costs $6,000. He has $500 for a down payment. He is deciding between a 2-year loan and a 3-year loan. Use the rates in the table to determine how much money he will save if he chooses the 2-year loan instead of the 3-year loan. Round to the nearest cent.

Time (years)	Simple Interest Rate (%)
1	1.5
2	2.25
3	3.0

12. Evelyn is buying a motorcycle that costs $14,000. The tax rate is 6.75%, and she plans to make a down payment of $1,500. The sales tax is added before the down payment is applied. She is considering a three-year loan. What will be her monthly payments? Round to the nearest cent.

Time (years)	Simple Interest Rate (%)
1	3.25
3	6.75
5	8.5

13. MP **Persevere with Problems** Suppose Serena invests $2,500 for 3 years and 6 months and earns $328.13. What was the rate of interest? Explain how you solved.

14. A student stated that an interest rate could not be less than 1%. Do you agree? Why or why not?

15. MP **Justify Conclusions** Suppose you earn 2% on $1,000 for 2 years. Explain how the simple interest is affected if the rate is doubled. What happens if the time is doubled?

16. Name a principal and interest rate where the amount of simple interest earned in 10 years would be $200. Justify your answer.

Commission and Fees

I Can... use proportional relationships to find the amount of commission earned on sales and the amount of fees for certain services.

What Vocabulary Will You Learn?
commission
fee

Learn Commission and Fees

The following are four ways in which people are paid.

Hourly	These employees are paid a set amount each hour, such as \$15 per hour.
Salary	These employees are paid a set amount, no matter how many hours they work, such as \$45,000 per year.
Commission Only	These employees are paid only a percent of the amount that they sell.
Salary plus Commission	These employees are paid a small base salary, plus a percent of the amount that they sell.

Some employees, such as realtors, car salesmen, and stockbrokers, are paid a percent of the amount that they sell. This payment is called a **commission**.

If you pay a commission to a person or business, you are paying a **fee**. A fee is a payment for a service. It can be a fixed amount, a percent of the charge, or both.

A realtor may earn 5% commission on the sale of a home. Suppose the realtor sells a house for \$140,000. You can use equivalent ratios, and the percentage written as a rate per 100, to determine the amount of commission. Let *c* represent the amount of commission.

amount of commission → ; price of house →
$$\frac{c}{140{,}000} = \frac{5}{100} \text{ } \} \text{ Percent}$$

$$\frac{7{,}000}{140{,}000} = \frac{5}{100} \quad (\times 1{,}400)$$

Because 100 × 1,400 is 140,000, multiply 5 by 1,400 to find *c*.

The amount of commission is \$7,000. This means the real estate agent will be paid \$7,000 for selling a \$140,000 home.

🌐 Example 1 Find Commission

Angie works in a jewelry store and earns a 6.25% commission on every piece of jewelry she sells.

How much commission does she earn for selling a ring that costs $1,300?

Method 1 Use ratio reasoning.

Write a proportion and solve using ratio reasoning. Let c represent the amount of commission.

$$\begin{array}{l}\text{amount of commission} \longrightarrow \\ \text{cost of ring} \longrightarrow\end{array} \left.\dfrac{c}{1{,}300} = \dfrac{6.25}{100}\right\} \text{Percent}$$

$$\dfrac{c}{1{,}300} = \dfrac{6.25}{100} \quad$$ Because 100 × 13 = 1,300, multiply 6.25 by 13 to find the value of c.

$$\dfrac{81.25}{1{,}300} = \dfrac{6.25}{100} \quad \begin{array}{l}6.25 \times 13 = 81.25,\\ \text{so, } c = 81.25.\end{array}$$

Method 2 Use properties of operations.

$$\dfrac{c}{1{,}300} = \dfrac{6.25}{100} \qquad \begin{array}{l}\text{Write the proportion. Let } c\\ \text{represent the amount.}\end{array}$$

$$\dfrac{c}{1{,}300} = 0.0625 \qquad \begin{array}{l}\text{Divide 6.25 by 100. A one-step}\\ \text{equation results.}\end{array}$$

$$1{,}300 \cdot \left(\dfrac{c}{1{,}300}\right) = (0.0625) \cdot 1{,}300 \qquad \text{Multiplication Property of Equality}$$

$$c = 81.25 \qquad \text{Simplify.}$$

So, using either method, the amount of commission is $81.25.

Check

Nathan sells jewelry and earns a 9.75% commission on every piece he sells. How much commission would he earn on a bracelet that sold for $220? Use any strategy.

(Show your work here)

🌐 **Go Online** You can complete an Extra Example online.

💭 Think About It!

What is a good estimate for the solution? Explain how you calculated the estimate.

💬 Talk About It!

How does your answer compare to your estimate?

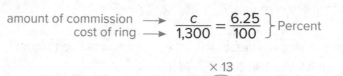

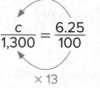

Example 2 Find the Amount of Sales

Caleb needs to earn $1,200 each month to cover his living expenses. He earns an 8% commission on everything he sells.

What is the minimum amount he needs to sell in order to earn $1,200?

Write a proportion. Then solve using ratio reasoning. Let a represent the amount he needs to sell.

amount of commission → 1,200; amount of sales → a

$$\frac{1{,}200}{a} = \frac{8}{100} \quad \text{Percent}$$

$$\frac{1{,}200}{a} = \frac{8}{100} \quad (\times 150)$$

Because $8 \times 150 = 1{,}200$, multiply 100 by 150 to find the value of a.

$$\frac{1{,}200}{15{,}000} = \frac{8}{100}$$

$100 \times 150 = 15{,}000$, so, $a = 15{,}000$.

So, Caleb needs to sell $15,000 to earn $1,200 in commission.

Check

Quentin wants to earn at least $945 this month in commission. What is the minimum amount he needs to sell if he earns a 5.25% commission?

Show your work here

Go Online You can complete an Extra Example online.

Pause and Reflect

Did you make any errors when completing the Check exercise? What can you do to make sure you don't repeat the errors in the future?

Record your observations here

Think About It!

Do you need to find the part, the percent, or the whole?

Talk About It!

How do you know your answer is reasonable?

Example 3 Fees

Sybrina bought some paper supplies from an online retailer. The retailer charges a $6.95 shipping fee or 8% of the total purchase, whichever is greater.

If Sybrina's total purchase is $85.50, how much shipping will she pay?

Step 1 Find the percent of the total purchase.

Because the shipping fee may be 8% of the total purchase, find 8% of $85.50.

Write a proportion and solve using the properties of operations. Let *x* represent the potential shipping fee.

$\frac{x}{85.50} = \frac{8}{100}$	Write the proportion.
$\frac{x}{85.50} = 0.08$	Divide 8 by 100.
$85.50 \cdot \left(\frac{x}{85.50}\right) = (0.08) \cdot 85.50$	Multiplication Property of Equality
$x = 6.84$	Simplify.

Step 2 Compare the two fees.

The 8% shipping fee of $6.84 is less than the $6.95 shipping fee. Sybrina must pay the greater amount.

So, Sybrina will pay $6.95 in shipping fees.

Check

Tickets for a concert are sold by online ticket resellers. One company charges a fee of 4% of the ticket price. Another company charges a flat fee of $2.50. The ticket you want to buy costs $60.50. If you buy the ticket online, which is the lesser fee?

Show your work here

Think About It!
What is a good estimate for an 8% shipping fee? Explain how you calculated the estimate.

Talk About It!
How can you solve the problem another way?

Go Online You can complete an Extra Example online.

Apply Personal Finance

David is starting a new job in sales. He can choose to earn only a 10% commission on sales each month, or earn a monthly base salary of $1,500 with a 3% commission on sales over $7,500. Which pay method would earn him more money if he has an average sales of $16,000 each month?

1 What is the task?

Make sure you understand exactly what question to answer or problem to solve. You may want to read the problem three times. Discuss these questions with a partner.

First Time Describe the context of the problem, in your own words.
Second Time What mathematics do you see in the problem?
Third Time What are you wondering about?

2 How can you approach the task? What strategies can you use?

Record your observations here

Talk About It!

How do you determine a 3% commission on sales over $7,500?

3 What is your solution?

Use your strategy to solve the problem.

Show your work here

4 How can you show your solution is reasonable?

Write About It! Write an argument that can be used to defend your solution.

Check

Kelly has a job in sales that pays only a commission of 24% of her total monthly sales. She is offered a new job with a monthly salary of $3,750 with a commission of 9.5% on her total monthly sales over $12,500. If she estimates her total monthly sales will average $20,000, which statement is correct?

(A) The new job will pay more because she would earn $4,462.50, and her current job only pays $4,220.

(B) The new job will pay more because she would earn $4,742.75, and her current job only pays $4,220.

(C) Her current job will pay more because she will earn $4,800, and the new job would only pay $4,462.50.

(D) Her current job will pay more because she will earn $4,800, and the new job would only pay $4,742.75.

Show your work here

Go Online You can complete an Extra Example online.

Pause and Reflect

How are *parts* and *wholes* represented in fees and commissions? Give examples to support your answer.

Record your observations here

Name ________________ Period ________ Date ________

Practice

Go Online You can complete your homework online.

Solve each problem. Use any strategy, such as a bar diagram, ratio table, or division.

1. Mrs. Hollern works in a jewelry store and earns an 8.5% commission on every piece of jewelry she sells. How much commission would she earn for selling a necklace that costs $4,600? **(Example 1)**

2. Booker's father sells computer software and earns a 4.25% commission on every software package he sells. How much commission would he earn on a software package that sold for $15,725? Round to the nearest cent. **(Example 1)**

3. Chase wants to earn at least $900 this month in commission. What is the minimum amount he needs to sell in order to earn $900 if he earns a 3.3% commission on everything he sells? Round to the nearest dollar. **(Example 2)**

4. Sophia needs to earn at least $1,800 each month to cover her living expenses. What is the minimum amount she needs to sell in order to earn at least $1,800 if she earns a 10.75% commission on everything she sells? Round to the nearest dollar. **(Example 2)**

5. Mrs. Jackson needs to earn $3,000 a month. How much does she need to sell if she earns 15% commission on everything she sells? **(Example 2)**

6. Raymond purchases concert tickets online for $43.50. There is a 3% processing fee. What is the total cost of the tickets? Round to the nearest cent. **(Example 3)**

7. The table shows the shipping fee options for two online marketplaces. If you sell a smart phone online for $115, which is the lesser fee? **(Example 3)**

Online Market	Shipping Fees
Deals	Flat fee of $12
Sell UR Stuff	11% of selling price

Test Practice

8. **Multiple Choice** Mai bought some party supplies from an online store. The shop charges a $7.95 shipping fee or 6.25% of the total purchase, whichever is greater. Suppose Mai's total purchase is $128. How much shipping will she pay?

Ⓐ $7.95

Ⓑ $8.00

Ⓒ $15.95

Ⓓ $16.00

Apply

9. Addison is considering two sales jobs. The job offers are shown in the table. She estimates her total monthly sales will average $18,000. Which job should Addison take if she wants to earn more money each month? How much more will she earn?

	Offer
Job 1	A commission of 24% of her total monthly sales.
Job 2	Salary of $4,000 with a commission of 12.25% on her total monthly sales over $14,500.

10. The table shows Frank and his brother's monthly salaries. One week, Frank and Daniel each have sales of $9,500. Who earned more money that week? How much more?

	Offer
Daniel	A commission of 15% on his weekly sales.
Frank	A weekly salary of $450 plus an 18% commission on his weekly sales over $4,000.

11. MP **Find the Error** A student is finding a 6.5% commission on $525.08 worth of sales. Find the student's mistake and correct it. Let x represent the amount of commission.

$$\frac{x}{525.08} = \frac{65}{100}$$

$$\frac{x}{525.08} = 0.65$$

$$x = 341.30$$

12. Create Write and solve a real-world problem in which you find the commission.

13. Determine if the following statement is *true* or *false*. Write an argument that can be used to defend your solution.

A commission can be less than 1%.

14. MP **Persevere with Problems** Natalie is a real estate agent and earns a 3% commission on all homes she sells. She estimates that she will earn $8,250 on the next house she sells. She actually earns $9,210. What was Natalie's estimate for the selling price of home? What was the actual selling price of the home?

Percent Error

I Can... use proportional relationships to solve percent error problems.

Explore Percent Error

Online Activity You will use Web Sketchpad to explore percent error.

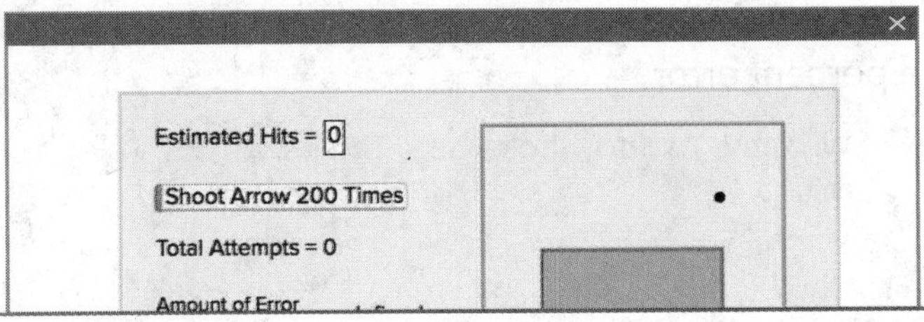

Learn Percent Error

Percent error is a ratio, written as a percent, that compares the inaccuracy of an estimate, or amount of error, to the actual amount.

The **amount of error** is the positive difference between the estimate and the actual amount. To find the amount of error, subtract the lesser amount from the greater amount.

Suppose 800 people are estimated to attend the high school football game. The actual attendance was 1,000 people. You can use a bar diagram to represent and find the percent of error.

Draw a bar to represent the actual attendance, 1,000 people. Because this is the whole, label the length of the bar 100%.

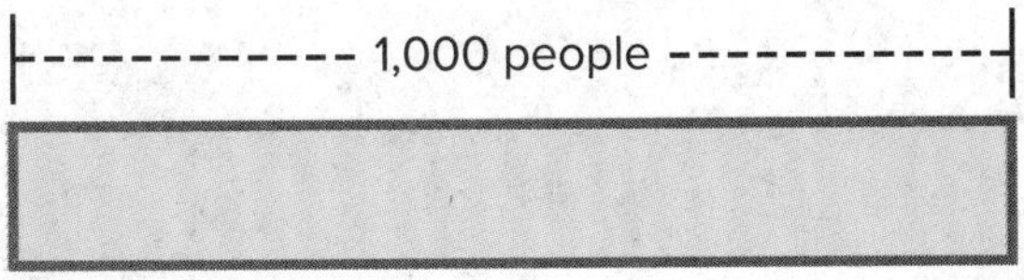

The amount of error is 1,000 − 800, or 200 people. Because 1,000 ÷ 200 = 5, divide the bar into 5 equal-size sections of 200.

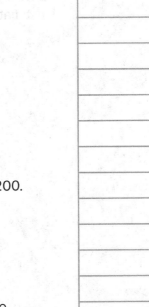

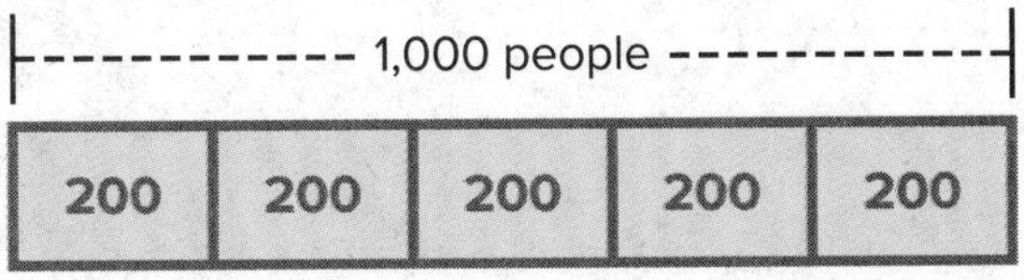

Each section represents 20% of the whole, 200 people. So, the percent error is 20%.

What Vocabulary Will You Learn?
amount of error
percent error

Talk About It!
Why do you think it is important to subtract the lesser amount from the greater amount when finding the amount of error?

Your Notes

Think About It!

Why might a bar diagram not be advantageous to help solve this problem?

Talk About It!

Suppose it actually took the contractor 19.5 hours. What part of solving the problem would stay the same? What part would change?

Example 1 Percent Error

A contractor estimates that it will take him 16 hours to complete a home improvement project. It actually takes him 12.5 hours.

What is the percent error of the contractor's estimate?

Step 1 Identify the part and the whole.

actual amount = 12.5	This is the whole.
estimated amount = 16	This is the whole plus the part.
amount of error = 16 − 12.5, or 3.5	This is the part.

Step 2 Find the percent error.

$\frac{\text{part}}{\text{whole}} = \frac{3.5}{12.5}$	Write the part-to-whole ratio. The part is 3.5. The whole is 12.5.
$= 0.28$	Divide.
$= \frac{28}{100}$	Write an equivalent ratio, as a rate per 100.
$= 28\%$	Definition of percent

So, the percent error is 28%.

Check

The Ramirez family was going on a trip. Their GPS system estimated that it would take them 4.25 hours to reach their destination. It actually took 5 hours because of stops. Find the percent error of the estimated time.

Show your work here

Go Online You can complete an Extra Example online.

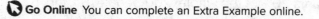

Apply Sports

A school newspaper estimates that their basketball team will win 23 out of 25 games for the season. After 10 games, they have won 8. If the team continues winning at this rate, what will be the percent error of the newspaper's estimate once the season is over?

1 What is the task?

Make sure you understand exactly what question to answer or problem to solve. You may want to read the problem three times. Discuss these questions with a partner.

First Time Describe the context of the problem, in your own words.
Second Time What mathematics do you see in the problem?
Third Time What are you wondering about?

2 How can you approach the task? What strategies can you use?

Record your observations here

3 What is your solution?

Use your strategy to solve the problem.

Show your work here

4 How can you show that your solution is reasonable?

Write About It! Write an argument that can be used to defend your solution.

Talk About It!
How can you predict the actual number of wins the team has at the end of the season?

Check

It is predicted that a softball team will win 35 out of their 50 games for their summer season. After 20 games, they have won 16. If the team continues to win at this rate, what will be the percent error of the prediction once the season is over?

Show your work here

Go Online You can complete an Extra Example online.

Pause and Reflect

Explain how ratio and proportional reasoning can be used to solve problems involving percent.

Record your observations here

Math History Minute

Marjorie Lee Browne (1914–1979) was one of the first African-American women to receive a doctorate degree in mathematics. While she was the head of the mathematics department at North Carolina College, she wrote a grant for $60,000 to IBM (International Business Machines) so that the university could have its first computer.

Name ______ Period ______ Date ______

Practice

Go Online You can complete your homework online.

Solve each problem.

1. Doug estimates that his soccer team will win 7 games this year. The team actually wins 10 games. What is the percent error of Doug's estimate? Round the answer to the nearest tenth percent, if necessary. **(Example 1)**

2. A mayor estimates that 4,000 people will attend the first day of the county fair. A total of 8,400 people actually attend the first day of the fair. What is the percent error of the mayor's estimate? Round the answer to the nearest tenth percent, if necessary. **(Example 1)**

3. Maya estimates that the wait time for her favorite roller coaster is 35 minutes. The actual wait time is 55.5 minutes. What is the percent error of Maya's estimate? Round the answer to the nearest tenth of a percent, if necessary. **(Example 1)**

4. Oliver estimates the weight of his cat to be 16 pounds. The actual weight of his cat is 14.25 pounds. What is the percent error of Oliver's estimate rounded to the nearest tenth of a percent? **(Example 1)**

5. A jar of marbles should contain 100 marbles. The jar actually has 99 marbles. What is the percent error to the nearest hundredth of a percent? **(Example 1)**

6. A cyclist estimates that he will bike 80 miles this week. He actually bikes 75.5 miles. What is the percent error of the cyclist's estimate rounded to the nearest hundredth of a percent? **(Example 1)**

7. The table shows the predicted and actual amount of snow for a local city. What is the percent error for the amount of snowfall? Round the answer to the nearest tenth of a percent if necessary. **(Example 1)**

	Snowfall (inches)
Predicted	6.75
Actual	10.25

Test Practice

8. Multiple Choice Jin's mother estimates there is a half gallon of fruit punch remaining in the container. There is actually 72 ounces of fruit punch in the container. What is the percent error of Jin's mother's estimate, rounded to the nearest tenth?

Ⓐ 11.1%

Ⓑ 36%

Ⓒ 88.9%

Ⓓ 144%

Apply

9. A school newspaper estimates that their academic team will win 25 out of 30 matches for the season. After 15 matches, they have won 12. If the team continues winning at this rate, what will be the percent error of the newspaper's estimate once the season is over? Round to the nearest percent.

10. A toy company that makes bubbles fills its 8-ounce bottles using a machine. To check that the machine fills the bottles with the proper amount, the company randomly checks bottles off the assembly line. A bottle passes inspection if the percent error of the amount is 2% or less. What is the range of values that a bottle could contain to pass inspection? Round to the nearest hundredth.

11. MP **Find the Error** A student is finding the percent error for an estimated length of 22 inches with an actual length of 25 inches. Find the student's mistake and correct it.

25 in. − 22 in. = 3 in.

$\frac{3}{25} = 0.12$

$= 0.12\%$

12. A student population of 1,200 was estimated to increase by 15% in the next five years. The population actually increased by 20%. Find the estimated and actual student populations and describe the percent error.

13. MP **Make an Argument** Make an argument for why you cannot find the percent error when the actual value is 0. Explain.

14. MP **Use a Counterexample** Determine if the following statement is *true* or *false*. If false, provide a counterexample.

Percent error can never be greater than 100%.

Review

Foldables Use your Foldable to help review the module.

Rate Yourself!

Complete the chart at the beginning of the module by placing a checkmark in each row that corresponds with how much you know about each topic after completing this module.

Write about one thing you learned.

____

____

____

____

Write about a question you still have.

____

____

____

____

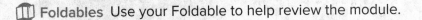

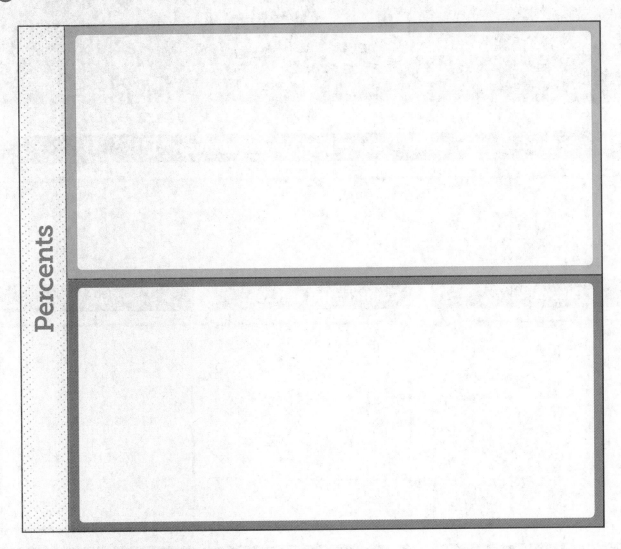

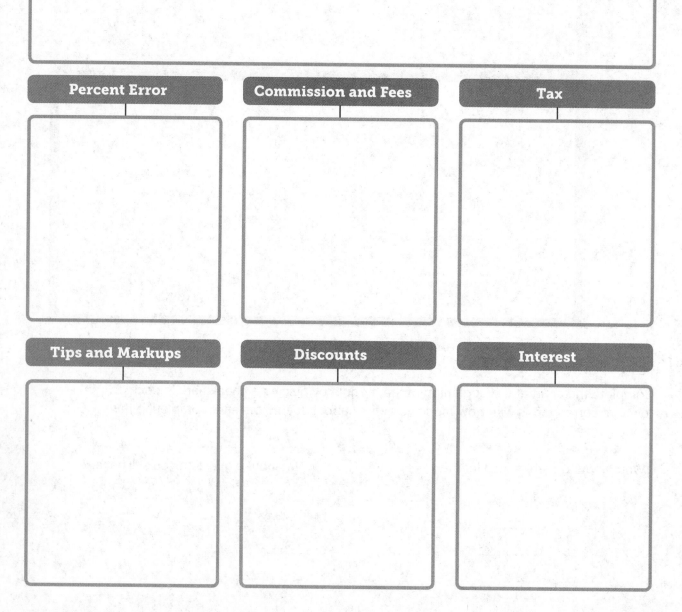

Reflect on the Module

Use what you learned about percents to complete the graphic organizer.

Essential Question

How can percent describe the change of a quantity?

How do you find the percent of change between two amounts?

Percent Error

Commission and Fees

Tax

Tips and Markups

Discounts

Interest

Name ______________________ Period __________ Date __________

Test Practice

1. **Open Response** An ice cream shop had 316 customers the first weekend it opened. The second weekend it had 468. What is the percent of increase in the number of customers? Round to the nearest percent. **(Lesson 1)**

2. **Open Response** Cynthia is training to run the 100-yard dash. Each week she records her best 100-yard dash time so that she can track her improvement. The table shows her best times after each of the first 3 weeks of training. **(Lesson 1)**

Week	Time (s)
1	14.0
2	13.5
3	13.1

A. Find the percent of change from Week 1 to Week 2, and from Week 2 to Week 3. Round to the nearest tenth percent if necessary.

B. How much greater is the percent of change between the first two weeks than between the second and third weeks? Round to the nearest tenth percent.

3. **Multiple Choice** The wholesale cost of a pair of sandals at a shoe store is $22.60. The markup for the sandals is 45%. What is the selling price of the sandals? **(Lesson 3)**

Ⓐ $10.17
Ⓑ $32.77
Ⓒ $41.15
Ⓓ $54.90

4. **Equation Editor** Julia wants to purchase a pair of headphones that are on sale for $42.79. The sales tax rate in her county is 6%. Calculate the amount of sales tax Julia will pay on the purchase. Round to the nearest cent if necessary. **(Lesson 2)**

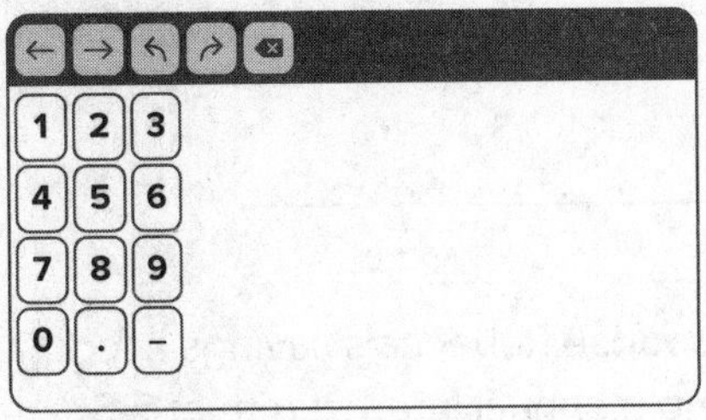

5. **Table Item** For parties of 6 or more people at a certain restaurant, a tip of 18% is automatically included on the bill.

Select *yes* or *no* to indicate whether or not each of the following tip amounts is correct for an 18% gratuity on each subtotal. **(Lesson 3)**

	yes	no
subtotal: $129.50 tip: $23.27		
subtotal: $98.40 tip: $17.71		
subtotal: $142.17 tip: $22.75		

6. **Multiple Choice** During a sale at a furniture store, sofas were marked down 30%. You have a coupon for an additional 15% off. What is the final price of a sofa that originally cost $550? **(Lesson 4)**

Ⓐ $302.50
Ⓑ $327.25
Ⓒ $396.00
Ⓓ $275.00

7. Equation Editor A set of curtains normally sells for $58.99. Roberta used a coupon good for 20% off the regular price. To the nearest cent, how much did Roberta pay for the curtains before sales tax? **(Lesson 4)**

8. Open Response Rebecca's savings account pays a simple annual interest rate of 2.5%. Suppose she deposits $4,280 in the account and makes no additional deposits or withdrawals for 4 years. What will the total value of the account be after 4 years? **(Lesson 5)**

9. Multiselect There are 118 different elements in the periodic table. While giving a presentation, Clark says that there are 128 different elements. Select each true statement. **(Lesson 7)**

- ☐ There are actually 118 elements.
- ☐ The amount of error in Clark's estimate is 10.
- ☐ The amount of error in Clark's estimate is 118.
- ☐ To the nearest tenth percent, the percent error in Clark's estimate is 8.5%.
- ☐ To the nearest tenth percent, the percent error in Clark's estimate is 10%.

10. Equation Editor Manny is a realtor and earns 3% commission on the sale of a house. If the house sells for $155,000, how much commission would he earn? **(Lesson 6)**

11. Open Response Mr. Gomez earns a 5.4% commission on his total sales for the month. If he wants to earn at least $3,500 in commission this month, what is the minimum amount he needs to sell? Round to the nearest dollar. Explain how you found your answer. **(Lesson 6)**

12. Multiple Choice Jake is a factory manager for a company that produces batteries. Occasionally, some are found to be defective. Based on previous results, Jake estimates that 12 batteries will be defective for a certain month. However, 10 were actually defective. What is the percent error of Jake's estimate? **(Lesson 7)**

Ⓐ 2%

Ⓑ 16%

Ⓒ 1.6%

Ⓓ 20%

Module 3

Operations with Integers

Essential Question

How are operations with integers related to operations with whole numbers?

What Will You Learn?

Place a checkmark (✓) in each row that corresponds with how much you already know about each topic **before** starting this module.

KEY

□ — I don't know. ◇ — I've heard of it. ☆ — I know it!

	Before			After		
	□	◇	☆	□	◇	☆
adding integers						
subtracting integers						
finding the distance between two integers on a number line						
multiplying integers						
dividing integers						
simplifying expressions involving integers using the order of operations						
evaluating algebraic expressions involving integers						

Foldables Cut out the Foldable and tape it to the Module Review at the end of the module. You can use the Foldable throughout the module as you learn about operations with integers.

What Vocabulary Will You Learn?

Check the box next to each vocabulary term that you may already know.

- ☐ absolute value
- ☐ Additive Inverse Property
- ☐ additive inverses
- ☐ Distributive Property
- ☐ Multiplicative Identity Property
- ☐ Multiplicative Property of Zero
- ☐ opposites
- ☐ order of operations

Are You Ready?

Study the Quick Review to see if you are ready to start this module. Then complete the Quick Check.

Quick Review

Example 1

Graph whole numbers on a number line.

Graph the number 5 on a number line.

Draw a dot at 5.

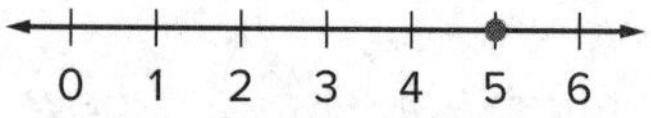

Example 2

Add whole numbers.

Add $4 + 9 + 16$.

$4 + 9 + 16$

$= 4 + 16 + 9$ Use the Commutative Property to reorder the addends.

$= 20 + 9$ Add 4 and 16.

$= 29$ Add 20 and 9.

Quick Check

1. Graph the number 2 on a number line.

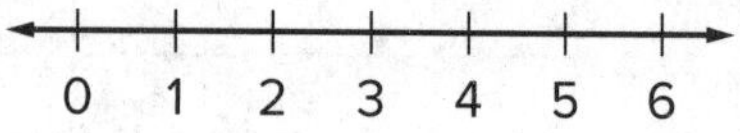

2. Justin downloaded 8 apps to his smartphone on Friday, 5 apps on Saturday, and 2 apps on Sunday. How many apps did he download for the 3 days?

How Did You Do?

Which exercises did you answer correctly in the Quick Check? Shade those exercise numbers at the right.

① ②

Add Integers

I Can... use different methods, including algebra tiles, number lines, or absolute value, to add integers.

What Vocabulary Will You Learn?
additive inverse
Additive Inverse Property
opposites

Explore Use Algebra Tiles to Add Integers

Online Activity You will use algebra tiles to model addition of integers, and make a conjecture about the sign of the sum of two integers.

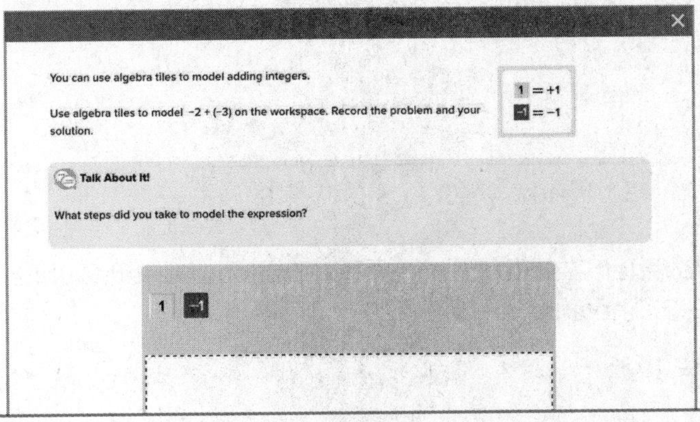

Learn Add Integers with the Same Sign

To add two integers with the same sign, you can use a horizontal or vertical number line.

The equation $-3 + (-4) = -7$ is modeled on the horizontal number line. Start at zero. Move left three units to model the negative integer, −3. Then, move left four units to model adding the negative integer −4. The sum is −7.

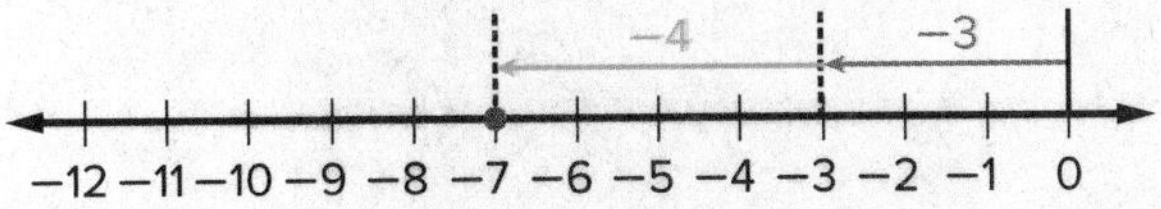

The equation $-5 + (-4) = -9$ is modeled on the vertical number line. Start at zero. Move down five units to model the negative integer −5. Then move down four units to model the negative integer −4. The sum is −9.

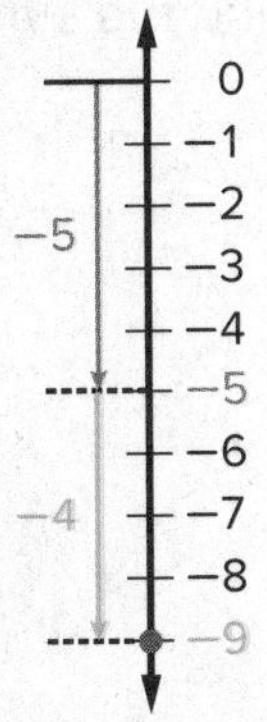

(continued on next page)

Your Notes

Talk About It!

How does a number line help show that the sum of two negative numbers will always be negative?

Talk About It!

Compare Method 1 and Method 2. Give an example of a situation where Method 1 might be more advantageous.

The number lines illustrate the rules for adding two integers with the same sign. To add two integers with the same sign, add their absolute values. The sum is:

- ______ if both integers are positive
- ______ if both integers are negative

Example 1 Add Integers with the Same Sign

Find −7 + (−2).

Method 1 Use a number line.

Go Online You can use the Web Sketchpad number line.

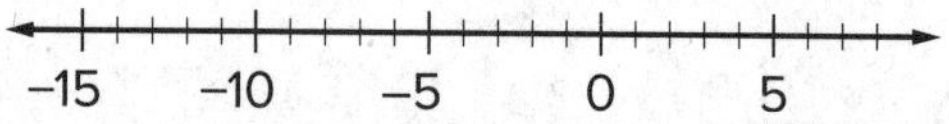

Start at zero. Move left 7 units to model −7. Then move left 2 units to model adding −2. The sum is −9.

So, −7 + (−2) = −9.

Method 2 Use the absolute value.

Because the integers have the same sign, find the sum of the absolute values.

$|-7| = \square$ and $|-2| = \square$

The sum of their absolute values is 7 + 2 or 9.

Because both integers are negative, the sum will be negative.

So, the solution to −7 + (−2) is −9.

Check

Find the sum of −5 + (−11).

Show your work here

Go Online You can complete an Extra Example online.

Example 2 Add Integers with the Same Sign

Allie borrowed $139 from her parents to purchase an ebook reader. During the first month, she purchased $47 in apps, games, and movies which was added to the amount she already owed her parents.

What integer represents the amount of money Allie had at the end of the first month?

The addition expression $-139 + (-47)$ represents the amount of money Allie had at the end of the month.

Because the integers have the same sign, find the sum of the absolute values.

$|-139| =$ ☐ and $|-47| =$ ☐

The sum of their absolute values is $139 + 47$ or ________.

Because both integers are negative, the sum will be ________.

So, the amount of money Allie has is $-\$186$.

Check

A contestant on a game show has −1,500 points. He loses another 750 points. What is his new score?

Show your work here

Go Online You can complete an Extra Example online.

Think About It!

Does a positive or negative integer represent borrowing and spending money?

Talk About It!

What does the value −$186 mean in the problem?

Learn Find Additive Inverses

The integers 4 and −4 are opposites. **Opposites** have the same absolute value but different signs. Two integers that are opposites are called **additive inverses** and their sum is zero.

The **Additive Inverse Property** can be used to find additive inverses.

Words	Example	Algebra
The sum of any number and its additive inverse is zero.	$4 + (-4) = 0$	$a + (-a) = 0$

Complete the table showing integers and their additive inverses.

Integer	Additive Inverse	Sum
−1	1	
2		0
	−3	0
−4		0
	5	0

Talk About It!

Which number is its own additive inverse? Explain.

Pause and Reflect

Are you ready to move on to the next Example? If yes, what have you learned that you think will help you? If no, what questions do you still have? How can you get those questions answered?

Record your observations here

Example 3 Find Additive Inverses

A hiking trail begins at an elevation of 150 feet above sea level. It leads down to the shore of the ocean, which has an elevation of 0 feet above sea level.

What integer represents the change in the elevation of the trail from beginning to end?

The trail begins at a positive height above the shore. To reach sea level at 0 feet, the change in elevation would have to be equal to the additive inverse of 150 feet. The additive inverse of 150 feet is −150 feet.

So, the integer that represents the change in elevation of the trail from beginning to end is ________.

Think About It!

Does the hiking trail ascend or descend?

Talk About It!

What do you notice about the signs of a pair of additive inverses?

Check

A puffin is flying at 29 feet above sea level. What is the elevation, in feet, that it will have to fly to reach sea level?

Show your work here

Go Online You can complete an Extra Example online.

Learn Add Integers with Different Signs

To add two integers with different signs, you can use a horizontal or vertical number line.

The horizontal number line models the equation $5 + (-3) = 2$. Start at zero. Move right five units to model the positive integer 5. Then move left three units to model adding the negative integer −3. The sum is 2.

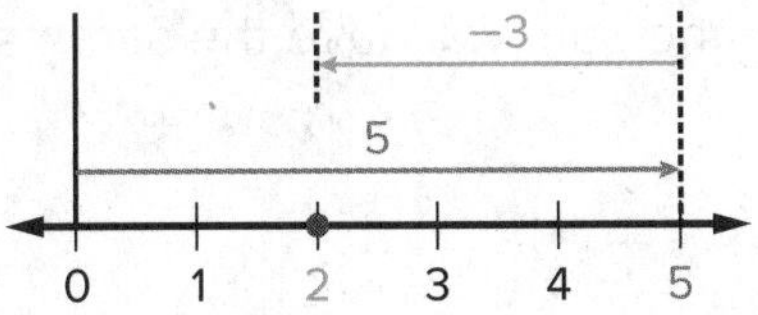

Predict how you can use a vertical number line to add the integers. Turn the page to check your prediction.

(continued on next page)

Talk About It!

Give an example of adding integers with different signs. Does your example reinforce the statements about the sign of a sum?

The vertical number line models the equation $-7 + 1 = -6$. Start at zero. Move down seven units to model the negative integer -7. Then move up one unit to model the positive integer 1. The sum is -6.

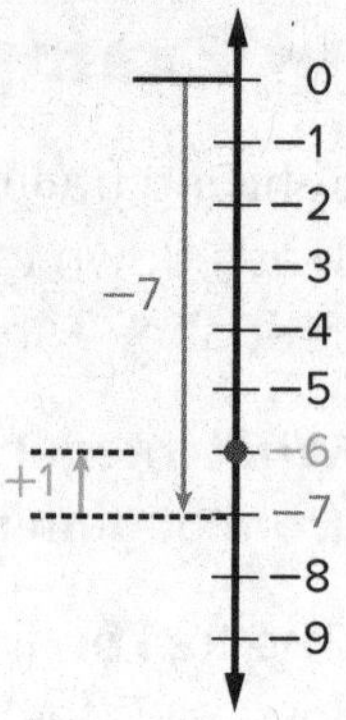

The number lines illustrate the rules for adding two integers with different signs. To add integers with different signs, subtract their absolute values. The sum is:

- positive if the positive integer's absolute value is greater
- negative if the negative integer's absolute value is greater

Example 4 Add Integers with Different Signs

Find $11 + (-4)$.

Method 1 Use a number line.

Go Online You can use the Web Sketchpad number line.

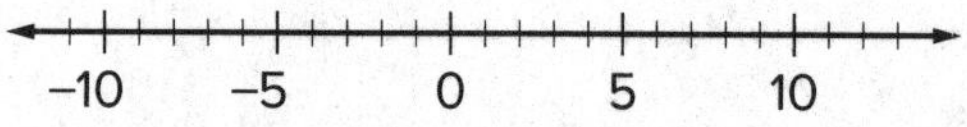

Start at zero. Move right 11 units to model 11. Then move left 4 units to model adding -4. The sum is 7.

So, $11 + (-4) =$ ☐.

Method 2 Use the absolute value.

Because the integers have different signs, find the difference of the absolute values.

$|11| =$ ☐ and $|-4| =$ ☐

The difference in their absolute values is $11 - 4$ or 7.

Because $|11| > |-4|$, the sum will have the same sign as ________.

So, $11 + (-4) = 7$.

Talk About It!

How could you change the second addend so that the sum is negative?

Check

Find the sum of $10 + (-22)$.

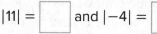

Show your work here

Go Online You can complete an Extra Example online.

Example 5 Add Integers with Different Signs

A whale swam at a depth of 275 feet below the surface of the ocean. After 10 minutes, it rose 194 feet.

What integer represents the location of the whale, relative to the ocean's surface, after 10 minutes? Interpret the integer within the context of the problem.

Because the integers have different signs, find the difference of the absolute values.

$|-275| =$ ☐ and $|194| =$ ☐

The difference of their absolute values is 275 − 194 or ________.

Because $|-275| > |194|$, the sum will have the same sign as ________.

So, the integer that describes the whale's location, in feet, is −81. This means the whale is 81 feet below the ocean's surface.

Talk About It!

Why would an answer of a positive integer not make sense in the context of the problem?

Talk About It!

Can you think of a change that could be made to the problem where a positive integer answer would make sense?

Check

At 5:00 P.M., a thermometer shows an outside temperature of 2°F. Then, over the next three hours, the temperature drops 11°F. Which integer represents the thermometer reading, in degrees, at 8:00 P.M.?

Show your work here

Go Online You can complete an Extra Example online.

Pause and Reflect

When you first saw this Example, what was your reaction? Did you think you could solve the problem? Did what you already know help you solve the problem?

Record your observations here

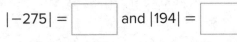

Think About It!

Can you group the negative addends together to add those first? What property allows you to do that?

Talk About It!

Compare and contrast Method 1 and Method 2.

Example 6 Add Three or More Integers

Find −26 + 74 + (−14).

Method 1 Add the numbers in order.

$-26 + 74 + (-14)$	Write the expression.
$= \square + (-14)$	Add −26 + 74.
$= \square$	Add 48 + (−14).

Method 2 Group like signs together.

$-26 + 74 + (-14)$	Write the expression.
$= -26 + (\square) + \square$	Use the Commutative Property.
$= \square + 74$	Add −26 + (−14).
$= \square$	Add.

So, the sum of −26 + 74 + (−14) is 34.

Check

Find −14 + 8 + (−6).

Show your work here

Go Online You can complete an Extra Example online.

Example 7 Add Three or More Integers

A roller coaster starts at point *A*, 43 feet above the ground. It ascends 55 feet, descends 80 feet, then ascends 110 feet to point *B*.

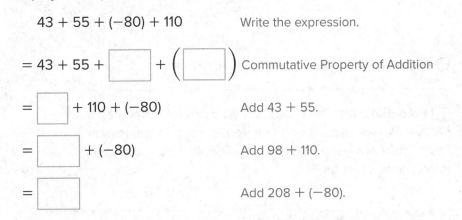

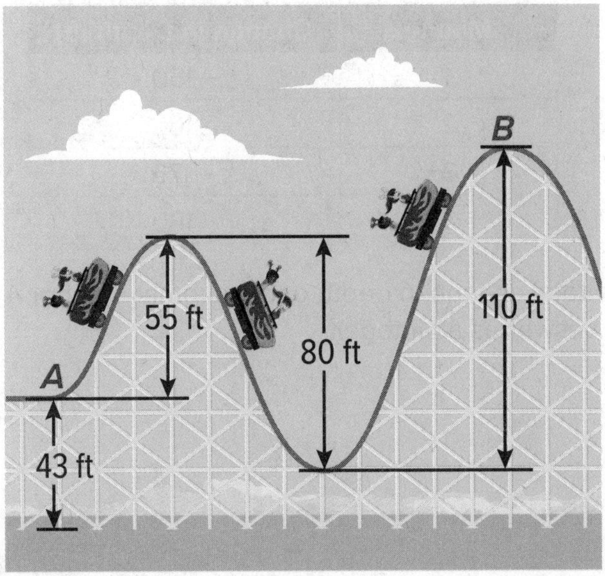

What is the height of the roller coaster at point *B* in relation to the ground?

Ascending can be represented by a positive integer and descending can be represented by a negative integer. So, the addition expression 43 + 55 + (−80) + 110 models the situation.

Simplify the expression.

43 + 55 + (−80) + 110	Write the expression.
= 43 + 55 + ☐ + (☐)	Commutative Property of Addition
= ☐ + 110 + (−80)	Add 43 + 55.
= ☐ + (−80)	Add 98 + 110.
= ☐	Add 208 + (−80).

So, point *B* is 128 feet above the ground.

Think About It!

What addition expression could represent this problem?

Talk About It!

Describe another way you can solve the problem.

Check

An unmanned submarine is doing tests along the ocean bottom. The following table shows the change in the depth of the submarine each minute for 4 minutes.

Minute	Change in Depth (ft)
1	−150
2	162
3	−175
4	−180

What is the total change in depth of the submarine after 4 minutes? Write your answer as an integer.

Show your work here

Go Online You can complete an Extra Example online.

Foldables It's time to update your Foldable, located in the Module Review, based on what you learned in this lesson. If you haven't already assembled your Foldable, you can find the instructions on page FL1.

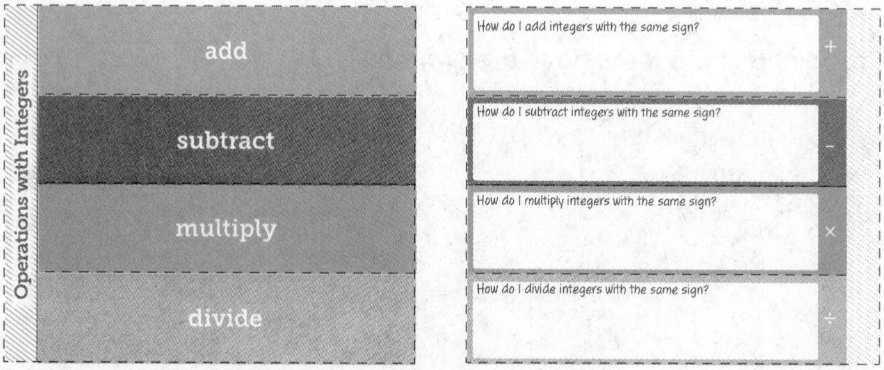
Operations with Integers
add
subtract
multiply
divide
How do I add integers with the same sign?
+
How do I subtract integers with the same sign?
−
How do I multiply integers with the same sign?
×
How do I divide integers with the same sign?
÷

Name ____________ Period ______ Date ______

Practice

Go Online You can complete your homework online.

Add. (Examples 1, 4, and 6)

1. $-3 + (-8)$

2. $-11 + (-13)$

3. $9 + (-35)$

4. $-28 + 14$

5. $-22 + (-10) + 15$

6. $18 + (-12) + 5$

7. Roger owes his father $15. He borrows another $25 from him. What integer represents the balance that he owes his father? **(Example 2)**

8. A football team lost 14 yards on their first play then lost another 7 yards on the next play. What integer represents the total change in yards for the two plays? **(Example 2)**

9. Kwan's beginning account balance was $20. His ending balance is $0. What integer represents the change in his account balance from beginning to end? **(Example 3)**

10. Lucy's dog lost 6 pounds. How much weight does her dog need to gain in order to have a net change of 0 pounds? **(Example 3)**

11. The table shows Jewel's scores for the first 9 holes and the second 9 holes of her game of golf. What integer represents her score for the entire game? **(Example 5)**

Holes	Score
1–9	3 over par
10–18	4 under par

12. At 4:00 A.M., the outside temperature was −28°F. By 4:00 P.M. that same day, it rose 38 degrees. What integer represents the temperature at 4:00 P.M.? **(Example 5)**

13. In 20 seconds, a roller coaster goes up a 100-meter hill, then down 72 meters, and then back up a 48-meter rise. How much higher or lower from the start of the ride is the coaster after the 20 seconds? **(Example 7)**

Test Practice

14. **Open Response** Joe opened a bank account with $80. He then withdrew $35 and deposited $115. What is his account balance after these transactions?

Apply

15. The table shows the transactions for one week for Tasha and Jamal. Who has the greater account balance at the end of the week? How much greater?

Transactions	Tasha	Jamal
Initial Deposit	\$250	\$200
Withdrawals	\$20	\$60
Deposits	\$65	\$135
Debit Card Purchases	\$46	\$27

16. A hot air balloon rises 340 feet into the air. Then it descends 130 feet, goes up 80 feet, and then down another 45 feet. How many feet will the balloon need to travel to return to the ground? Represent this amount as an integer. Explain.

17. What value of x would result in the numerical value of zero for each expression?

a. $-10 + 11 + x$

b. $7 + x + (-10)$

c. $x + 1 + (-1)$

18. MP **Find a Counterexample** Patrick stated that the sum of a positive integer and a negative integer is always negative. Find a counterexample that illustrates why this statement is not true.

19. Explain how you know that the sum of 2, 3, and -2 is positive without computing.

20. Create Write and solve a real-world problem where you add three integers and the sum is negative.

Subtract Integers

I Can... use different methods, including algebra tiles, number lines, or the additive inverse, to subtract integers.

Explore Use Algebra Tiles to Subtract Integers

Online Activity You will use algebra tiles to model subtraction of integers, and draw conclusions about the sign of the difference of the two integers.

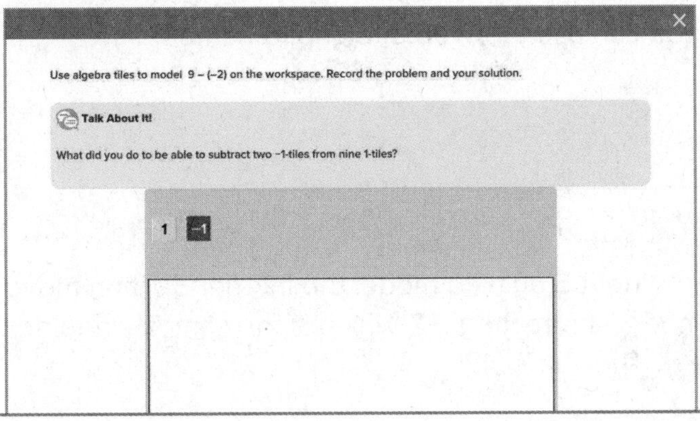

Use algebra tiles to model 9 – (–2) on the workspace. Record the problem and your solution.
Talk About It!
What did you do to be able to subtract two –1-tiles from nine 1-tiles?
1
–1

Learn Subtract Integers

To subtract integers, you can use a horizontal or vertical number line.

The horizontal number line models the equation $4 - 9 = -5$. Start at zero. Move right four units to model the integer 4. Then move left nine units to model subtracting 9. The difference is -5.

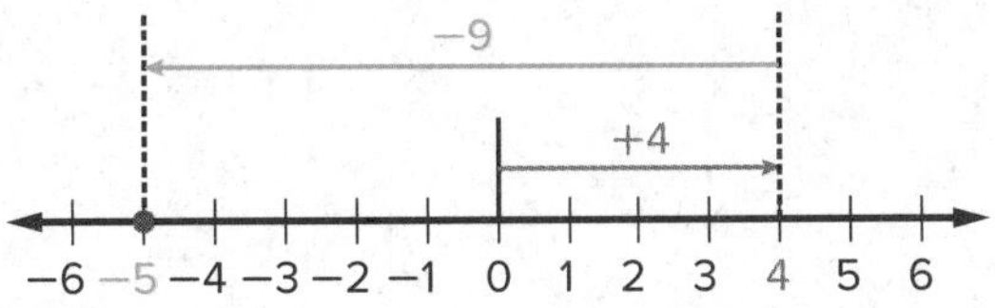

The vertical number line models the equation $4 - 9 = -5$. Start at zero. Move up four units to model the integer 4. Then move down nine units to model subtracting 9. The difference is -5.

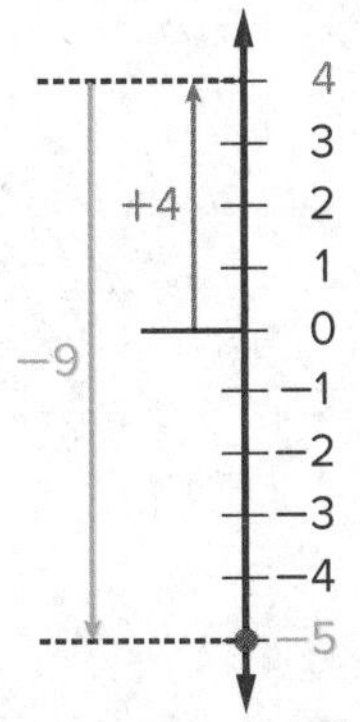

Talk About It!
The Commutative Property is true for addition. For example, $7 + 2 = 2 + 7$. Is the Commutative Property true for subtraction? Does $7 - 2 = 2 - 7$? Explain your reasoning using a number line.

(continued on next page)

Your Notes

Think About It!

Will you use a number line or will you add the additive inverse to solve this problem?

Talk About It!

Compare and contrast Method 1 and Method 2.

The number lines illustrate the rules for subtracting two integers.

Words	Symbols	Example
To subtract an integer, add the additive inverse of the integer.	$p - q = p + (-q)$	$4 - 9 = 4 + (-9) = -5$

Example 1 Subtract Integers

Find 5 – (–7).

Method 1 Use a number line.

Go Online You can use the Web Sketchpad number line.

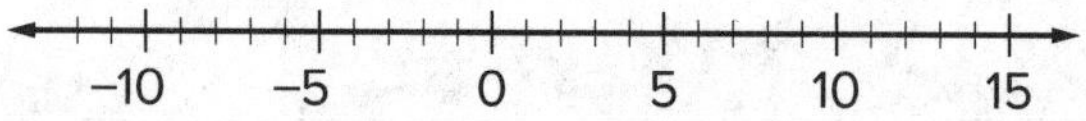

Start at zero. Move right 5 units to model the integer 5. Then move right 7 units to model subtracting –7, which is the same as adding the additive inverse, 7. The sum is 12.

So, $5 - (-7) = 12$.

Method 2 Use the additive inverse.

$5 - (-7) = 5 + 7$	To subtract –7, add the additive inverse of –7.
$= \square$	Add 5 + 7.

So, $5 - (-7) = 12$.

Check

Find 11 – (–15).

Show your work here

Go Online You can complete an Extra Example online.

Example 2 Subtract Integers

Find −24 − (−17).

$$-24 - (-17) = -24 + 17 \qquad \text{To subtract } -17, \text{ add its additive inverse.}$$
$$= -7 \qquad\qquad \text{Add.}$$

So, $-24 - (-17) = -7$.

Check

Find $-39 - (-24)$.

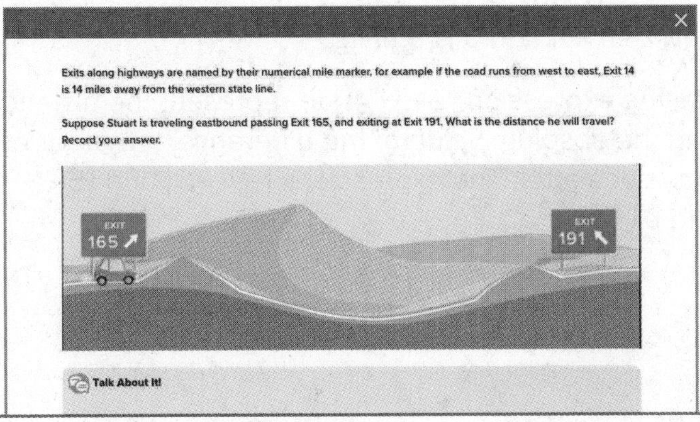

Think About It!
Predict the sign of the difference between the two integers.

Example 3 Subtract Expressions

Evaluate $x - y$ if $x = -23$ and $y = 19$.

$$x - y = -23 - 19 \qquad \text{Replace } x \text{ with } -23 \text{ and } y \text{ with } 19.$$
$$= -23 + (-19) \qquad \text{To subtract } 19, \text{ add its additive inverse.}$$
$$= -42 \qquad\qquad \text{Add } -23 + (-19).$$

So, when $x = -23$, and $y = 19$, $x - y = -42$.

Check

Evaluate $p - q$ if $p = -21$ and $q = 37$.

Talk About It!
Describe a situation where the difference between two numbers is greater than either number. Then explain why that happens.

Go Online You can complete an Extra Example online.

Explore Find Distance on a Number Line

Online Activity You will calculate distance traveled by using a number line to find the difference of the two integers.

Exits along highways are named by their numerical mile marker, for example if the road runs from west to east, Exit 14 is 14 miles away from the western state line.

Suppose Stuart is traveling eastbound passing Exit 165, and exiting at Exit 191. What is the distance he will travel? Record your answer.

EXIT 165
EXIT 191

Talk About It!

Learn Find the Distance Between Integers

Find the distance between −4 and 5.

Go Online Watch the animation to learn how to find the distance between two integers.

Method 1 Use a number line.

Step 1 Plot the integers on a number line. The animation shows two points at −4 and 5.

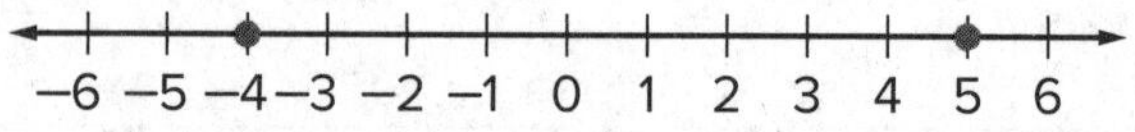

Step 2 Count the number of units between the two integers.

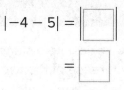

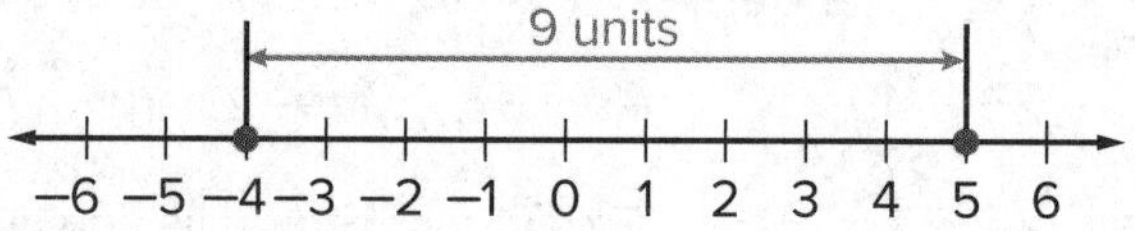

There are 9 units between −4 and 5.

Method 2 Use an expression.

The distance between two integers is equal to the absolute value of their difference.

distance = |difference of integers|

Step 1 Write an expression for the distance.

$$|-4 - 5|$$

Step 2 Simplify the expression.

$$|-4 - 5| = |\square|$$

$$= \square$$

The distance between −4 and 5 is 9 units.

You can also use the expression $|5-(-4)|$ to represent the distance. Because you find the absolute value of the difference, the order of the integers does not matter. The expressions $|-4-5|$ and $|5-(-4)|$ are both equal to 9.

Talk About It!

Why do we take the absolute value of the difference?

Example 4 Find the Distance Between Integers

Find the distance between −9 and 8.

Method 1 Use a number line.

Go Online You can use the Web Sketchpad number line.

Start at −9. Move right until you reach 8.

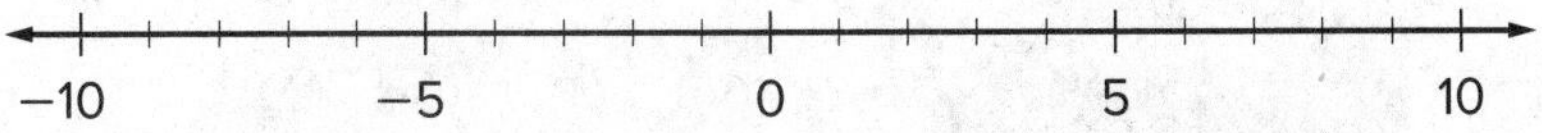

There are ________ units between −9 and 8.

Method 2 Use the absolute value.

To find the distance between integers, you can find the absolute value of their difference.

$|-9 - 8| = |-9 + (-8)|$ Add the additive inverse of 8.

$= |\square|$ or $\square$ Simplify.

So, the distance between −9 and 8 is 17 units.

Check

Find the distance between −5 and 9 on the number line.

Show your work here

Go Online You can complete an Extra Example online.

Pause and Reflect

When finding the distance between integers with different signs, which method would you choose to use? Explain.

Record your observations here

Think About It!

What subtraction expression could be used to find the distance?

Talk About It!

Compare and contrast the two methods.

Example 5 Find the Distance Between Integers

The highest point in California is Mount Whitney with an elevation of 14,494 feet. The lowest point is Death Valley with an elevation of −282 feet.

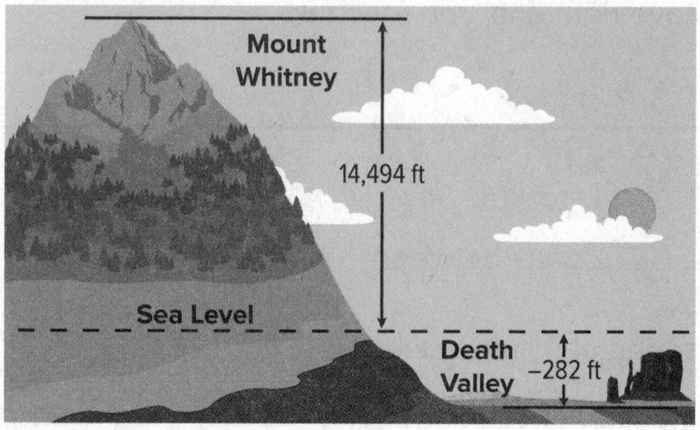

What is the distance between the height of Mount Whitney and the depth of Death Valley?

$\lvert 14{,}494 - (-282) \rvert = \lvert 14{,}494 + 282 \rvert$	To subtract −282, add its additive inverse.
$= \lvert 14{,}776 \rvert$	Add.
$= \square$	Find the absolute value.

So, the distance between the two points is 14,776 feet.

Check

The top of an iceberg is 55 feet above sea level, while the bottom is 385 feet below sea level. What is the distance between the top and bottom of the iceberg?

Show your work here

Think About It!
Will the distance be greater or less than 14,494 feet?

Talk About It!
Is it reasonable to have a negative answer? Why or why not?

Go Online You can complete an Extra Example online.

Apply The Solar System

The table shows the minimum and maximum temperatures on various celestial objects in the solar system.

Celestial Object	Minimum Temperature (°F)	Maximum Temperature (°F)
Moon	−387	253
Mars	−225	70
Mercury	−279	801
Venus	864	864

Scientists want to send a probe to study the celestial object with the greatest variation in temperature. To which celestial object should they send the probe?

1 What is the task?

Make sure you understand exactly what question to answer or problem to solve. You may want to read the problem three times. Discuss these questions with a partner.

First Time Describe the context of the problem, in your own words.
Second Time What mathematics do you see in the problem?
Third Time What are you wondering about?

2 How can you approach the task? What strategies can you use?

Record your observations here

3 What is your solution?

Use your strategy to solve the problem.

Show your work here

4 How can you show your solution is reasonable?

Write About It! Write an argument that can be used to defend your solution.

Go Online Watch the animation.

Talk About It!

On which celestial object from the table would it be most reasonable to live? Explain.

Check

The table shows the highest and lowest points of elevation, in relation to sea level, in four countries. Which country in this list has the greatest variation in elevation? the least?

Country	Highest Point (ft)	Lowest Point (ft)
Jordan	6,083	−1,404
United Kingdom	4,406	−13
Sweden	6,903	−8
Ireland	3,406	−10

Show your work here

Go Online You can complete an Extra Example online.

Foldables It's time to update your Foldable, located in the Module Review, based on what you learned in this lesson. If you haven't already assembled your Foldable, you can find the instructions on page FL1.

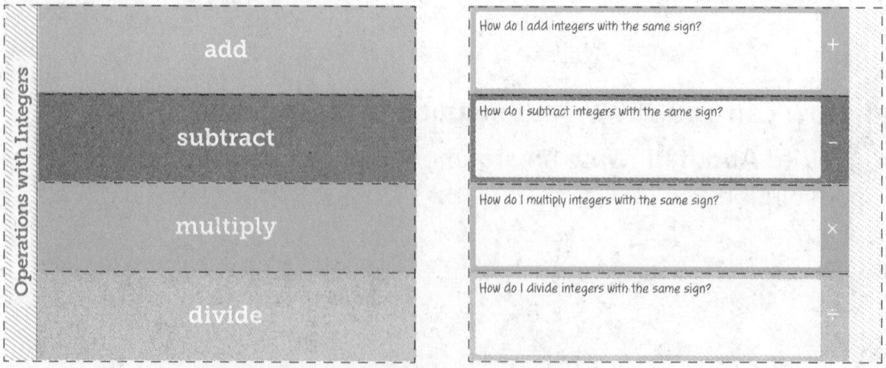
Operations with Integers
add
subtract
multiply
divide
How do I add integers with the same sign?
+
How do I subtract integers with the same sign?
−
How do I multiply integers with the same sign?
×
How do I divide integers with the same sign?
÷

Name ______________________ Period ________ Date ________

Practice

Go Online You can complete your homework online.

Subtract. **(Examples 1 and 2)**

1. $9 - (-2)$

2. $-20 - 10$

3. $13 - (-63)$

4. $28 - 14$

5. $-10 - 0$

6. $-33 - 33$

7. $-18 - (-12)$

8. $-28 - (-13)$

9. $-18 - (-40)$

10. Evaluate $a - b$ if $a = 10$ and $b = -7$. **(Example 3)**

11. Evaluate $x - y$ if $x = -11$ and $y = 26$. **(Example 3)**

12. Find the distance between −6 and 7 on a number line. **(Example 4)**

13. Find the distance between −14 and 5 on a number line. **(Example 4)**

14. The highest and lowest recorded temperatures for the state of Texas are 120° Fahrenheit and −23° Fahrenheit. Find the range of these extreme temperatures. **(Example 5)**

Test Practice

15. Open Response The table shows the starting and ending elevations of a hiking trail. How much greater is the elevation of the ending point than the starting point for the trail?

Point on Trail	Elevation
Starting Point	180 ft below sea level
Ending Point	260 ft above sea level

Apply

16. The table shows the maximum and minimum account balances for three college students for one month. Giovanni claimed that he had the least variation (from maximum to minimum) in his account balance that month. Is he correct? Write a mathematical argument to justify your solution.

Student	Maximum Balance ($)	Minimum Balance ($)
Jordan	145	−25
Giovanni	168	15
Elisa	152	−10

17. The table shows the record high and record low temperatures for certain U.S. states. Which state in the list had the greatest variation in temperature? the least?

State	Record High Temperature (°F)	Record Low Temperature (°F)
Alaska	100	−80
Idaho	118	−60
Nevada	125	−50
Utah	117	−69

18. MP **Use a Counterexample** Determine if each statement is *true* or *false*. If false, provide a counterexample.

a. Distance is always positive.

b. Change is always positive.

19. MP **Find the Error** A student is finding $4 - (-2)$. Find the student's mistake and correct it.

$$4 - (-2) = 4 - 2$$
$$= 2$$

20. Create Write a subtraction expression with a positive and negative integer whose difference is negative. Then find the difference.

21. If you subtract two negative integers, will the difference *always, sometimes,* or *never* be negative? Explain using examples to justify your solution.

Multiply Integers

I Can... use number lines and mathematical properties to multiply integers.

What Vocabulary Will You Learn?
additive inverse
Distributive Property
Multiplicative Identity Property
Multiplicative Property of Zero

Explore Use Algebra Tiles to Multiply Integers

Online Activity You will use algebra tiles to model integer multiplication, and make a conjecture about the sign of the product of the two integers.

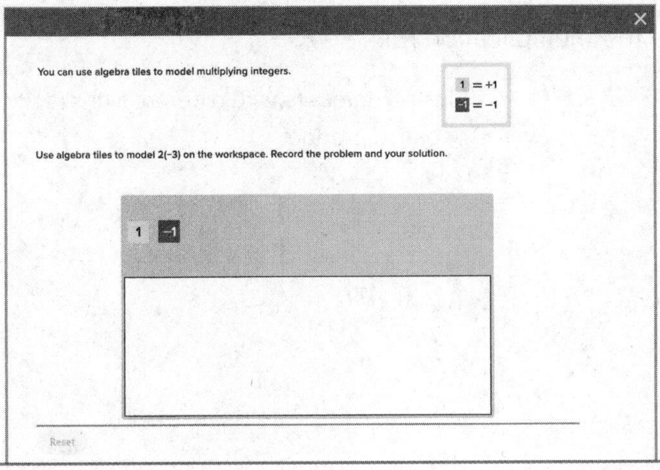

You can use algebra tiles to model multiplying integers.
1 = +1
−1 = −1
Use algebra tiles to model 2(−3) on the workspace. Record the problem and your solution.
1
−1
Reset

Learn Multiply Integers with Different Signs

Because multiplication is repeated addition, you can use a number line to show that 3(−6) means that −6 is used as an addend 3 times.

The number line models 3(−6).

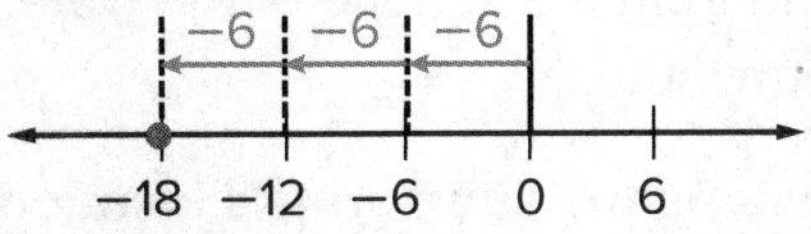

The number line illustrates the rule for multiplying integers with different signs.

Words	Examples
The product of two integers with different signs is negative.	$3(-6) = -18$ $-3(6) = -18$

Your Notes

Think About It!
Predict the sign of the two integers.

Talk About It!
How could you represent $-7(5)$ as an addition expression?

Think About It!
What multiplication expression could you use to solve Example 2?

Talk About It!
In the expression, why is 90 negative? Why would a positive integer not make sense as the product in this situation?

Example 1 Multiply Integers with Different Signs

Find $-7(5)$.

Method 1 Use a number line.

Using the Commutative Property of Multiplication, you can rewrite $-7(5)$ as $5(-7)$. Use a number line to show five groups of -7.

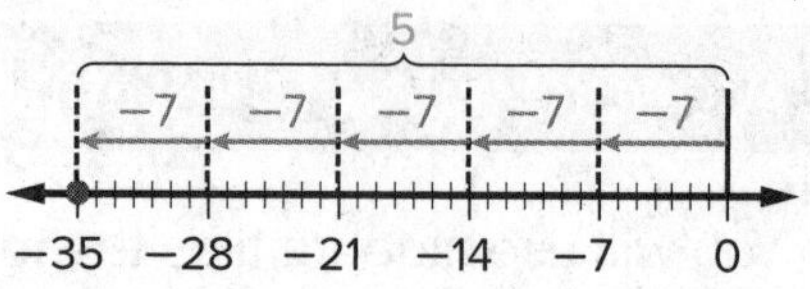

$-7(5) = \square$

Method 2 Use the multiplication rule.

$-7(5) = \square$ Multiply. Integers with different signs result in a negative product.

So, $-7(5) = -35$.

Check

Find $9(-13)$.

Show your work here

Go Online You can complete an Extra Example online.

Example 2 Multiply Integers with Different Signs

A submarine is diving from the surface of the water and descends at a rate of 90 feet per minute.

What integer represents the submarine's location, in feet, after 11 minutes?

The submarine descends at a rate of 90 feet per minute for 11 minutes. The expression $11(-90)$ represents the situation.

The signs of the integers are different. The product is negative.

$11(-90) = \square$

So, after 11 minutes, the location of the submarine will be 990 feet below the surface.

Check

A helicopter descends at a rate of 275 feet per minute. What integer represents the change in the helicopter's altitude, in feet, after 7 minutes?

Show your work here

Go Online You can complete an Extra Example online.

Learn Multiply Integers with the Same Sign

Go Online Watch the animation to learn how to multiply integers with the same sign.

The animation shows that you can use a pattern to show that the product of two negative numbers is positive. Notice that when you decrease the first factor by 1, the product increases by 3. You can continue the pattern to negative numbers.

$(2)(-3) = -6$

$(1)(-3) = -3$

$(0)(-3) = 0$

$(-1)(-3) = 3$

$(-2)(-3) = 6$

The pattern shows that the product of two negative numbers is positive.

When multiplying two integers with the same sign, such as 5 and 6, or -5 and -6, the sign of the product is always positive.

Words	Examples
The product of two integers with the same sign is positive.	$6(5) = 30$ $-6(-5) = 30$

Math History Minute

Negative numbers were not widely recognized by mathematicians until the 1800s, with a few exceptions. In 6th century India, negative numbers were introduced to represent debts and Indian mathematician **Brahmagupta (598–668)** stated rules for adding, subtracting, multiplying, and dividing negative numbers.

Example 3 Multiply Integers with the Same Sign

Find $-8(-9)$.

$-8(-9) = \square$ — The signs of the integers are the same. The product is positive.

So, the product of $-8(-9)$ is 72.

Check

Find $-5(-13)$.

Show your work here

Example 4 Multiply Integers with the Same Sign

Evaluate xy if $x = -14$ and $y = -7$.

$xy = -14(-7)$ — Replace x with -14 and y with -7.

$= \square$ — The signs of the integers are the same. The product is positive.

So, the value of the expression is 98.

Check

Evaluate pq if $p = -16$ and $q = -7$.

Show your work here

Go Online You can complete an Extra Example online.

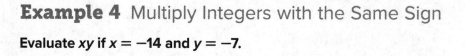

Example 5 Multiply Three or More Integers

Find −4(−7)(−2).

$-4(-7)(-2) = [-4(-7)](-2)$ Associative Property

$= \square(-2)$ Multiply (−4)(−7).

$= \square$ Multiply 28(−2).

So, the product of −4(−7)(−2) is −56.

Check

Find −3(−11)(−3).

Show your work here

Example 6 Multiply Three or More Integers

Evaluate ab^2c when $a = -5$, $b = 4$, and $c = -9$.

$ab^2c = -5(4)^2(-9)$ Substitute −5 for *a*, 4 for *b*, and −9 for *c*.

$= -5(\square)(-9)$ Multiply 4 • 4.

$= [-5(16)](-9)$ Associative Property

$= \square(-9)$ Multiply −5(16).

$= \square$ Multiply −80(−9).

So, the value of the expression is 720.

Check

Evaluate *pqrs* if $p = -7$, $q = 15$, $r = 1$, and $s = -2$.

Show your work here

Go Online You can complete an Extra Example online.

Think About It!

What do you notice about the expression?

Talk About It!

How can you tell, without multiplying, if the product will be positive or negative?

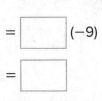

Learn Use Properties to Multiply Integers

Talk About It!

How can the properties of multiplication help you multiply integers?

In mathematics, properties can be used to justify statements you make while verifying or proving another statement. Some of the properties of mathematics are listed below.

Additive Inverse Property	$a + (-a) = 0$
Distributive Property	$a(b + c) = ab + ac$
Multiplicative Identity Property	$1 \cdot a = a$
Multiplicative Property of Zero	$a \cdot 0 = 0$

Go Online Watch the animation online to learn how to use properties of multiplication to multiply integers.

The animation shows that $2(-1) = -2$ using properties of multiplication, beginning with the true statement $2(0) = 0$.

$2(0) = 0$	Multiplicative Property of Zero
$2[1 + (-1)] = 0$	Additive Inverse Property
$2(1) + 2(-1) = 0$	Distributive Property
$2 + 2(-1) = 0$	Multiplicative Identity Property

In order for $2 + 2(-1)$ to equal 0, $2(-1)$ must equal -2, based on the Additive Inverse Property. This shows, using properties, that the product of two integers with different signs is negative.

You can also use properties to show that the product of two negative integers is positive.

Show that $(-2)(-1) = 2$ by writing the correct property for each step.

$0 = -2(0)$ ______________________

$0 = -2[1 + -1]$ ______________________

$0 = -2(1) + (-2)(-1)$ ______________________

$0 = -2 + (-2)(-1)$ ______________________

$2 = (-2)(-1)$ ______________________

Apply Agriculture

Two farms begin the year with $1,500 of extra savings each in case they lose money at the year's end. The table shows the amount of money each farm earned, or revenue, and the amount of money each farm spent, or expenses, for one month. If these results are consistent with the revenue and expenses for each month of the year, which farm will have enough savings to continue to operate for the whole year and how much will they have left over?

	Farm 1		Farm 2	
	Revenue ($)	Expenses ($)	Revenue ($)	Expenses ($)
Farm Supplies		134		211
Water		44		248
Maintenance		152		147
Potatoes Sold	308		476	

1 What is the task?

Make sure you understand exactly what question to answer or problem to solve. You may want to read the problem three times. Discuss these questions with a partner.

First Time Describe the context of the problem, in your own words.
Second Time What mathematics do you see in the problem?
Third Time What are you wondering about?

2 How can you approach the task? What strategies can you use?

Record your observations here

3 What is your solution?

Use your strategy to solve the problem.

Show your work here

4 How can you show your solution is reasonable?

Write About It! Write an argument that can be used to defend your solution.

Talk About It!

To the nearest integer, how much would each farm have to begin with in order for them both to have savings left over? Explain.

Check

James started a lawn care business in his community. He thinks that he should be able to operate his business for 20 weeks before it gets too cold. The table shows the expenses and revenue after one week.

Item	Expenses ($)	Revenue ($)
Gasoline	32	
Lawn Mower Maintenance	12	
Grass Seed and Weed Killer	17	
Income ($25 per lawn)		125

How much money will he earn in 20 weeks?

Show your work here

Go Online You can complete an Extra Example online.

Foldables It's time to update your Foldable, located in the Module Review, based on what you learned in this lesson. If you haven't already assembled your Foldable, you can find the instructions on page FL1.

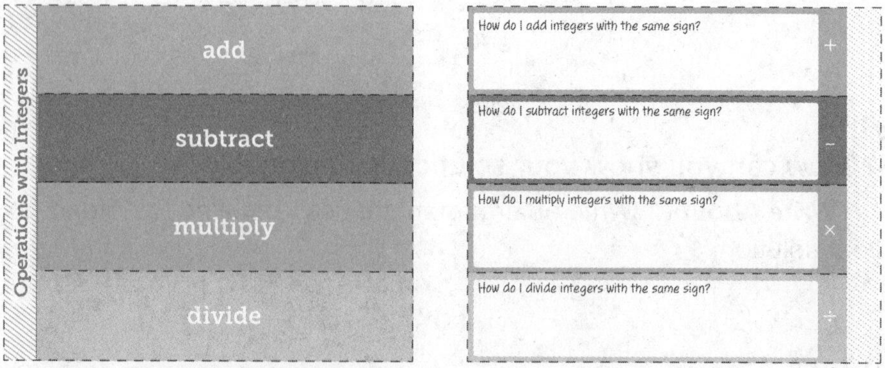
Operations with Integers
add
subtract
multiply
divide
How do I add integers with the same sign?
+
How do I subtract integers with the same sign?
−
How do I multiply integers with the same sign?
×
How do I divide integers with the same sign?
÷

Name ______________________ Period __________ Date __________

Practice

Go Online You can complete your homework online.

Multiply. (Examples 1, 3, and 5)

1. $4(-7)$

2. $-14(5)$

3. $9(-12)$

4. $-6(-8)$

5. $-10(-10)$

6. $-11(-13)$

7. $7(-5)(4)$

8. $(-8)(-7)(3)$

9. $-2(-12)(-8)$

10. Evaluate ab if $a = -16$ and $b = -5$. (Example 4)

11. Evaluate xy if $x = -10$ and $y = -7$. (Example 4)

12. Evaluate xyz^2 if $x = -2$, $y = 7$, and $z = -4$. (Example 6)

13. Evaluate a^2bc if $a = 3$, $b = -14$, and $c = -6$. (Example 6)

14. Mrs. Rockwell lost money on an investment at a rate of $4 per day. What is the change in her investment, due to the lost money, after 4 weeks? (Example 2)

Test Practice

15. Open Response The table shows the number of questions answered incorrectly by each player on a game show. If each missed question is worth −7 points, what is the change in Olive's score due to the incorrect questions?

Player	Incorrect Questions
Laura	8
Olive	9

Apply

16. Payton starts a lemonade stand for the summer. She thinks that she should be able to operate her business for 14 weeks. The table shows the expenses and revenue after 1 week. Based on this, how much money will Payton make during the 14 weeks?

Item	Expenses	Revenue
Cups	\$5	
Lemonade	\$6	
Ice	\$7	
Income		\$45

17. Rakim's goal is to have at least \$500 in his checking account at the end of the year. The table shows his activity for the month of January. Will Rakim make his goal if the month of January's activity is representative of how much he will save and spend each month? Explain.

Transaction	Amount
Debit Card Purchases	\$500
ATM Withdraws	\$750
Deposits	\$1,300

18. MP **Reason Inductively** The product of two integers is -24. The difference between the two integers is 14. The sum of the two integers is 10. What are the two integers?

19. MP **Identify Structure** Name the property illustrated by the following.

a. $-x \cdot 1 = -x$

b. $x \cdot (-y) = (-y) \cdot x$

20. If you multiply three negative integers, will the product *always, sometimes,* or *never* be negative. Explain.

21. MP **Identify Structure** Name all the values of x if $6|x| = 48$.

Divide Integers

I Can... use a related multiplication sentence to divide integers.

Explore Use Algebra Tiles to Divide Integers

Online Activity You will use algebra tiles to model integer division, and check solutions using multiplication.

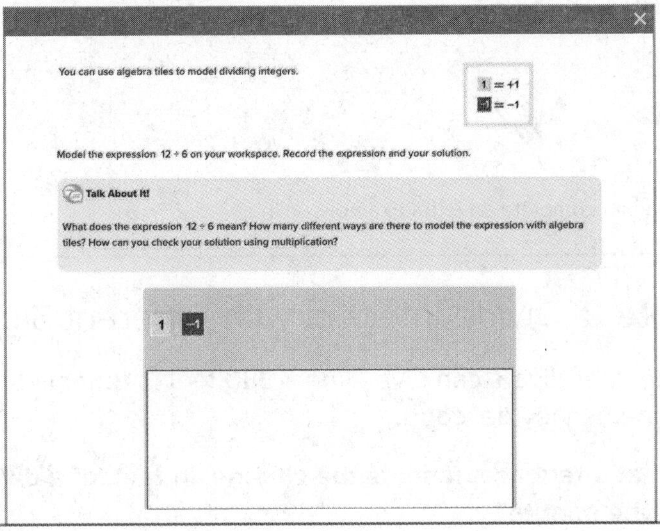

You can use algebra tiles to model dividing integers.
1 = +1
−1 = −1
Model the expression 12 ÷ 6 on your workspace. Record the expression and your solution.
Talk About It!
What does the expression 12 ÷ 6 mean? How many different ways are there to model the expression with algebra tiles? How can you check your solution using multiplication?
1
−1

Learn Divide Integers with Different Signs

Division is the inverse operation of multiplication, the same way that subtraction is the inverse of addition. You can find a quotient by using a related multiplication sentence.

$$36 \div (-3) = -12 \quad \rightarrow \quad -3 \cdot (-12) = 36$$

What number multiplied by −3 results in 36? ______

So, 36 ÷ (−3) is −12.

The table shows the rules for dividing integers with different signs.

Words	Examples
The quotient of two integers with different signs is negative.	$-36 \div 3 = -12$ $36 \div (-3) = -12$

Talk About It!

What do you think is the quotient of −36 ÷ 3? Explain your reasoning.

Your Notes

Think About It!

Predict the sign of the quotient.

Talk About It!

A friend stated that the quotient was positive. How can you use multiplication to show that the quotient is negative?

Talk About It!

What does the solution −95 represent in the context of the problem?

Example 1 Divide Integers with Different Signs

Find 90 ÷ (−10).

$90 \div (-10) = -9$ — The signs of the integers are different. The quotient is negative.

So, 90 ÷ (−10) = ________.

Check

Find 72 ÷ (−9).

Show your work here

Go Online You can complete an Extra Example online.

Example 2 Divide Integers with Different Signs

The best underwater divers can dive almost 380 feet in four minutes, using only a rope as they descend.

At this rate, what integer represents the change, in feet, of a diver's position after one minute?

Step 1 Identify the distance descended.

What integer represents the change, in feet, in the diver's position after four minutes? ☐

Step 2 Write the division expression.

☐ ÷ ☐ — Divide the total change in the diver's position by the number of minutes, 4.

Step 3 Divide −380 by 4 to find the change after one minute.

$-380 \div 4 =$ ☐ — The signs of the integers are different. The quotient is negative.

So, the integer −95 represents the change in the diver's position, in feet, after one minute.

Check

A neighbor is improving his dog's fitness level by working on agility training. Before beginning the training regiment, the dog weighed 65 pounds. After nine weeks of training, the dog weighed 47 pounds. What integer represents the change in weight that the dog averaged, in pounds per week, over the nine weeks?

Show your work here

Learn Divide Integers with the Same Sign

Division is the inverse operation of multiplication. You can find a quotient by finding a related multiplication sentence.

positive quotient

$-20 \div (-4) = 5 \quad \rightarrow \quad -4(\square) = -20$

same signs

The table shows the rules for dividing integers with the same sign.

Words	Examples
The quotient of two integers with the same sign is positive.	$20 \div 4 = 5$ $-20 \div (-4) = 5$

Example 3 Divide Integers with the Same Sign

Find $-30 \div (-6)$.

$-30 \div (-6) = \square$

The signs of the integers are the same. The quotient is positive.

So, $-30 \div (-6) = 5$.

Check

Find $-84 \div (-12)$.

Show your work here

Go Online You can complete an Extra Example online.

Talk About It!

How are division and multiplication of integers similar? How are they different?

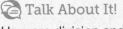

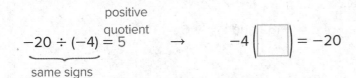

Example 4 Divide Integers with the Same Sign

Evaluate $\frac{y}{x}$ if $y = -155$ and $x = -5$.

$\frac{y}{x} = \frac{-155}{-5}$ Replace y with -155 and x with -5.

$= \square \div \square$ Write as a division sentence.

$= \square$ The signs of the integers are the same. The quotient is positive.

So, the value of the expression is 31.

Check

Evaluate $\frac{y}{x}$ if $y = -162$ and $x = -6$.

Show your work here

Go Online You can complete an Extra Example online.

Pause and Reflect

Compare and contrast how to predict the sign of the integer when adding or subtracting, and when multiplying or dividing.

Record your observations here

Apply Personal Finance

Natalie had $165 in her bank account at the beginning of the summer. Over the next 10 weeks, she worked at a summer camp and added $160 to her savings each week, while spending only $40 per week. Once she gets back to school, she plans to spend $105 per week. For how many weeks can she make withdrawals until her balance is $0?

1 What is the task?

Make sure you understand exactly what question to answer or problem to solve. You may want to read the problem three times. Discuss these questions with a partner.

First Time Describe the context of the problem, in your own words.
Second Time What mathematics do you see in the problem?
Third Time What are you wondering about?

2 How can you approach the task? What strategies can you use?

Record your observations here

3 What is your solution?

Use your strategy to solve the problem.

Show your work here

4 How can you show your solution is reasonable?

Write About It! Write an argument that can be used to defend your solution.

Talk About It!
If Natalie wants to be able to withdrawal $105 for 15 weeks, how much can she spend each week during the summer?

Check

You start a pool cleaning business in the neighborhood. You start your business with $1,000 of savings. The table shows the expenses and revenue after one week. At this rate, how many weeks will your savings last?

Item	Expenses	Revenue
Cleaning Chemicals	$31	
Brushes and Towels	$17	
Transportation	$10	
Income ($15 per pool)		$30
Flyers for Advertising	$12	

Show your work here

Go Online You can complete an Extra Example online.

Foldables It's time to update your Foldable, located in the Module Review, based on what you learned in this lesson. If you haven't already assembled your Foldable, you can find the instructions on page FL1.

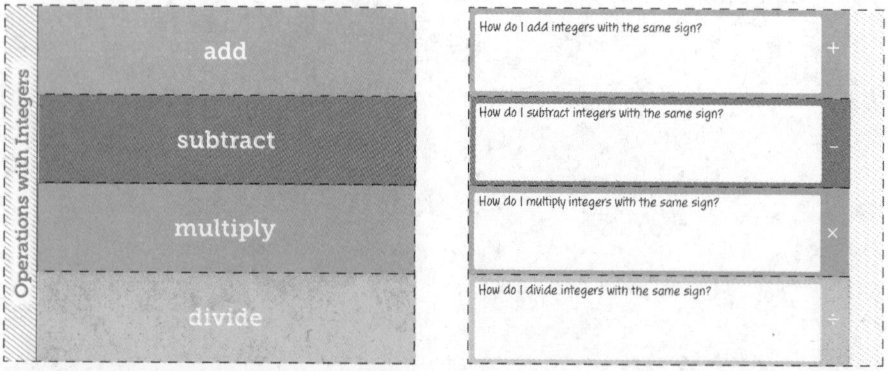
Operations with Integers
add
subtract
multiply
divide
How do I add integers with the same sign?
+
How do I subtract integers with the same sign?
−
How do I multiply integers with the same sign?
×
How do I divide integers with the same sign?
÷

Name ______ Period ______ Date ______

Practice

Go Online You can complete your homework online.

Divide. **(Examples 1 and 3)**

1. $22 \div (-2)$

2. $-110 \div 11$

3. $75 \div (-3)$

4. $-64 \div (-8)$

5. $-39 \div (-13)$

6. $-50 \div (-10)$

Evaluate each expression if $m = -32$, $n = 2$, and $p = -8$. **(Example 4)**

7. $\frac{m}{n}$

8. $\frac{m}{p}$

9. $\frac{p}{n}$

Evaluate each expression if $f = -15$, $g = 5$, and $h = -45$. **(Example 4)**

10. $\frac{f}{g}$

11. $\frac{h}{f}$

12. $\frac{h}{g}$

13. A submarine descends to a depth of 660 feet below the surface in 11 minutes. At this rate, what integer represents the change, in feet, of the submarine's position after one minute? **(Example 2)**

Test Practice

14. Equation Editor Aaron made 3 withdrawals last month. Each time, he withdrew the same amount. If Aaron withdrew a total of \$375, what integer represents the change in his account after the first withdrawal?

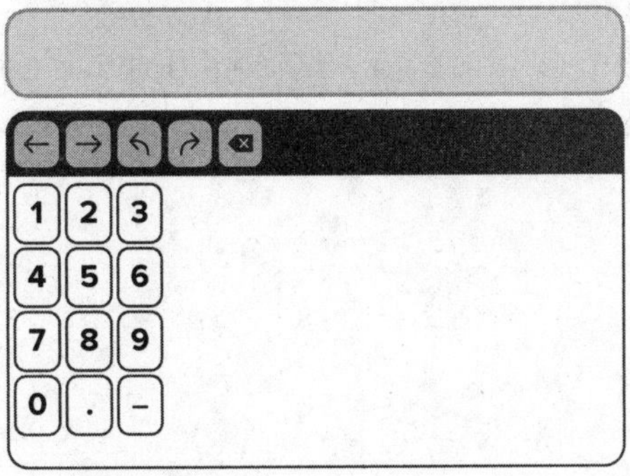

Apply

15. Over the summer, Paulo opens a dog-washing business and begins with \$32. The table shows how much he earns and spends each week. He works for 8 weeks washing dogs. When he starts back at school, he budgets \$60 to spend each week. How many weeks pass before he needs to wash more dogs?

Weekly Revenue and Expenses		
	Revenue	Expense
Sales	\$124	
Supplies		\$8

16. Nick had \$100 in his savings account. Over the next 6 months, he worked at a seasonal store, where each month he earned \$400 and spent \$250. He put the remaining amount in his savings account each month. Now that the job is over, he plans to spend \$200 per month. For how many months can he make withdrawals from his savings account until his balance is \$0?

17. MP **Make an Argument** The Associative Property holds true for multiplication because $(-3 \times 4) \times (-2) = -3 \times [4 \times (-2)]$. Does the Associative Property hold true for division of integers? Explain.

18. MP **Justify Conclusions** Is the following statement *true* or *false*? Justify your response.

If n is a negative integer, $\frac{n}{n} = -1$.

19. Create Write and solve a real-world problem in which you divide a positive and negative integer.

20. Write a division sentence that divides a negative integer by a positive integer. Then write a multiplication sentence that proves your division sentence is correct.

Apply Integer Operations

I Can... use the order of integer operations to evaluate expressions.

Example 1 Order of Integer Operations

Find −4(3) + (−7).

$-4(3) + (-7) = -12 + (-7)$ Multiply −4(3).

$= -19$ Add.

So, −4(3) + (−7) is −19.

Check

Find −5(−12) + (−15).

Show your work here

Talk About It!

Why was multiplication performed as the first operation in evaluating the expression?

Example 2 Order of Integer Operations

Find −4(−5)(−2) − (−8).

$-4(-5)(-2) - (-8) = 20(-2) - (-8)$ Multiply −4(−5).

$= -40 + 8$ Multiply 20(−2). Add the additive inverse of −8.

$= -32$ Add −40 + 8.

So, −4(−5)(−2) − (−8) is −32.

Check

Find 9(−1)(−8) − (−13).

Show your work here

Talk About It!

Could you have multiplied (−5)(−2) first? Explain your reasoning.

Go Online You can complete an Extra Example online.

Your Notes

Think About It!
How would you begin evaluating the expression?

Example 3 Order of Integer Operations

Evaluate $\frac{w}{xy} + y - z^3$ if $w = 36$, $x = -6$, $y = -1$, and $z = -2$.

$\frac{w}{xy} + y - z^3 = \frac{36}{(-6)(-1)} + (-1) - (-2)^3$	Substitute the values.
$= \frac{36}{(-6)(-1)} + (-1) - (-8)$	$(-2)^3 = (-2)(-2)(-2)$
$= \frac{36}{6} + (-1) - (-8)$	Multiply $(-6)(-1)$.
$= 6 + (-1) - (-8)$	Divide $\frac{36}{6}$.
$= 5 - (-8)$	Add $6 + (-1)$.
$= 5 + 8$	Add the additive inverse.
$= 13$	Add $5 + 8$.

So, the value of the expression is ☐.

Check

Evaluate $\frac{q}{rs} - (r \cdot p)$ if $q = 56$, $r = -4$, $s = 2$, and $p = 1$.

Show your work here

Example 4 Order of Integer Operations

The average temperature in January in Helsinki, Finland is about −5°C. **Use the expression $\frac{(9C + 160)}{5}$, where C is the temperature in degrees Celsius, to find the temperature in degrees Fahrenheit. Round to the nearest degree.**

$\frac{(9C + 160)}{5} = \frac{9(-5) + 160}{5}$	Replace C with −5.
$= \frac{-45 + 160}{5}$	Multiply 9(−5).
$= \frac{115}{5}$	Add −45 + 160.
$= 23$	Simplify.

So, the average temperature in January in Helsinki, Finland is about 23 degrees Fahrenheit.

Talk About It!
At what temperature do you think the measures in °F and °C are equal? Explain your reasoning.

Check

In a recent year, the average temperature during the month of June in Hall Beach, Canada was 32°F. Use the expression $\frac{5(F - 32)}{9}$, where F is the temperature in degrees Fahrenheit, to find the temperature in degrees Celsius. Round to the nearest degree.

Go Online You can complete an Extra Example online.

Name ________________ Period ________ Date ________

Practice

Go Online You can complete your homework online.

Evaluate each expression. (Examples 1 and 2)

1. $-5(6) + (-9)$

2. $\frac{-36}{9} + (-7)$

3. $-4(-8) + (-10)$

4. $2(-5)(-6) - (-12)$

5. $10(-3)(4) - (-15)$

6. $2\left(\frac{-80}{4}\right) - 14$

Evaluate each expression if $a = -2$, $b = 3$, $c = -12$, and $d = -4$. (Example 3)

7. $\frac{bd}{a} + c$

8. $\frac{ac}{b} - (a + d)$

9. $\frac{d^3}{a^2} - (c + b)$

Evaluate each expression if $m = -32$, $n = 2$, $p = -8$, and $r = 4$. (Example 3)

10. $\frac{pr}{n} + m$

11. $\frac{p^2}{m} - (np + r)$

12. $\frac{p^3}{r^2} - (m + np)$

13. The table gives the income and expenses of a small company for one year. Use the expression $\frac{I - E}{12}$ where I represents the total income and where E represents the total expenses, to find the average difference between the company's income and expenses each month. (Example 4)

	Amount ($)
Income	84,000
Expenses	86,400

Test Practice

14. **Open Response** Five years ago the population at Liberty Middle School was 1,600 students. This year the population is 1,250 students. Use the expression $\frac{N - P}{5}$ where N represents this year's population and where P represents the previous population to find the average change in population each year.

Apply

15. The table shows the extreme temperatures for different U.S. cities in degrees Fahrenheit. Use the expression $\frac{5(F-32)}{9}$, where F represents the temperature in degrees Fahrenheit to convert each temperature to degrees Celsius. Which city had the greatest difference in temperature extremes in degrees Celsius? Which city had the least? Round to the nearest degree. Explain.

City	Low Extreme (°F)	High Extreme (°F)
Chicago	−27	104
Nashville	−17	107
Oklahoma City	−8	110

16. The table shows the yearly initial and ending balance for each sibling in a family for their account balance with their parents. Use the expression $\frac{I-E}{12}$ where I represents the initial balance and where E represents the ending balance, to find the average difference between the initial balance and ending balance each month. Which sibling had the greatest monthly change?

Sibling	Initial Balance	Ending Balance
Alma	\$226	−\$50
Laurel	\$200	−\$64
Wes	\$290	\$2

17. MP **Find the Error** A student solved the problem shown below. Find the student's mistake and correct it.

Find $-3(-6)(-4) - (-9)$.

$$(18)(-4) - (-9) = (-72) - (-9)$$
$$= -81$$

18. Create Write and solve a real-world problem in which you perform more than one operation with integers.

19. When simplifying $5(-2)(9) - 3$, a student first subtracted 3 from 9. Is this the correct first step? Explain.

20. MP **Identify Structure** When simplifying $-7(3)(-10) + (-4)$, can you multiply 3 and -10 first and get the same result as multiplying -7 and 3 first? Explain.

Review

Foldables Use your Foldable to help review the module.

Operations with Integers
How do I add integers with different signs?
How do I subtract integers with different signs?
How do I multiply integers with different signs?
How do I divide integers with different signs?

Rate Yourself!

Complete the chart at the beginning of the module by placing a checkmark in each row that corresponds with how much you know about each topic after completing this module.

Write about one thing you learned.

Write about a question you still have.

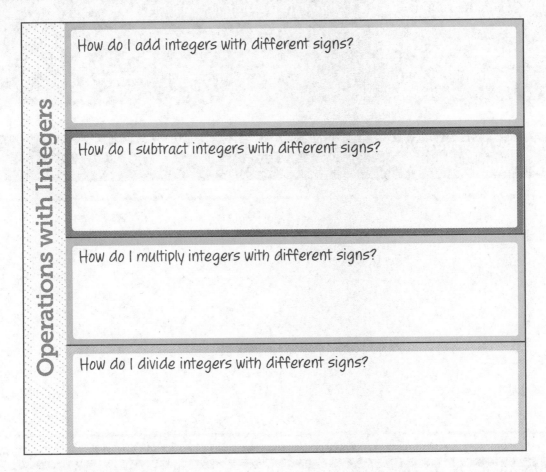

Reflect on the Module

Use what you learned about integers to complete the graphic organizer.

Essential Question

How are operations with integers related to operations with whole numbers?

How do you determine the sign of the sum or difference when you add and subtract integers?

How do you determine the sign of the product or quotient when you multiply and divide integers?

Name ______________________ Period __________ Date __________

Test Practice

1. Open Response The table shows Tammy's golf score with respect to par for 2 rounds of a tournament. **(Lesson 1)**

Round	Score (strokes to par)
1	+5
2	−2

What is her total score with respect to par?

2. Multiple Choice Terrell uses his debit card to buy \$54 worth of groceries. Later he withdraws \$25 from his checking account. What integer represents the total change in his account balance? **(Lesson 1)**

Ⓐ −\$29

Ⓑ −\$79

Ⓒ \$29

Ⓓ \$79

3. Open Response Christine is finding the sum 28 + (−15) + 22. **(Lesson 1)**

A. What property of addition could Christine use to write the integers in a different order? Explain why she might want to do this.

B. Find the sum of 28 + (−15) + 22.

4. Equation Editor The table shows the high temperature on the moon during the day and the overnight low temperature. **(Lesson 2)**

Time of Day	Temperature
Day	253°F
Night	−387°F

What is the range between the moon's minimum and maximum temperature in degrees Fahrenheit?

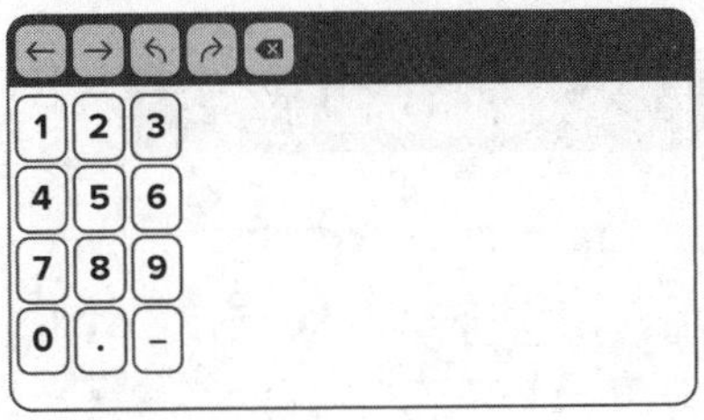

5. Multiple Choice What number sentence is modeled by the number line shown below? **(Lesson 2)**

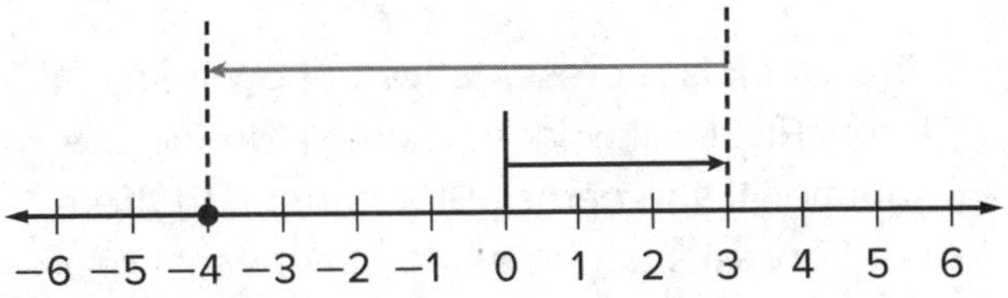

Ⓐ $7 - 7 = 0$

Ⓑ $-4 + 7 = 3$

Ⓒ $3 - 7 = -4$

Ⓓ $3 - 4 = -1$

6. Equation Editor Find ab^2c if $a = -3$, $b = -2$, and $c = 6$. **(Lesson 3)**

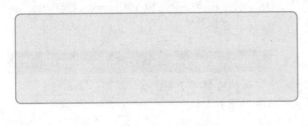

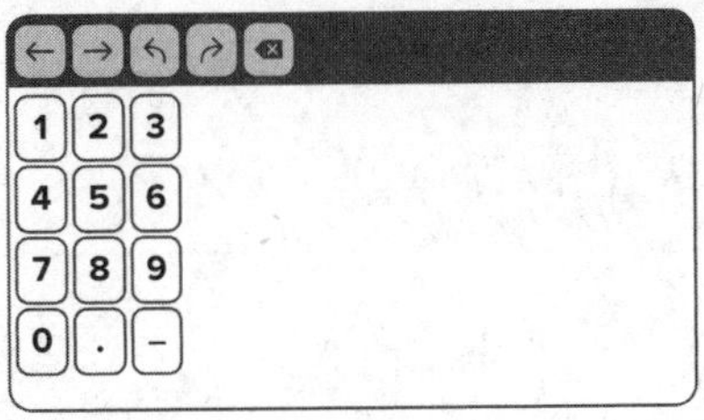

7. Equation Editor Find $-5(10)(-4)$. **(Lesson 3)**

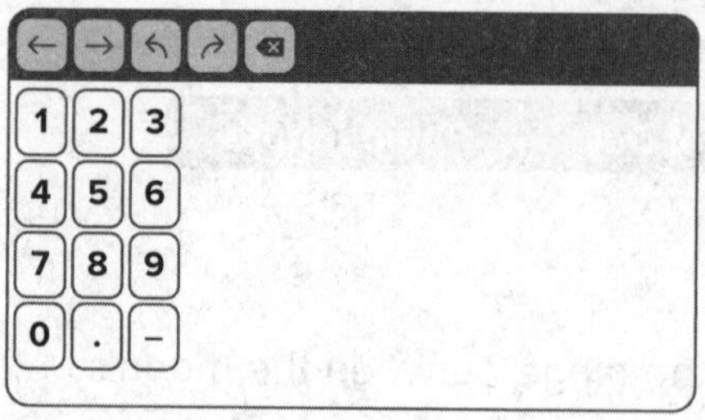

8. Open Response Ricki is working a part-time job this summer. The table shows how many hours she worked each day during her first week. **(Lesson 3)**

Day	Hours Worked
Monday	4
Tuesday	0
Wednesday	5
Thursday	5
Friday	0
Saturday	3
Sunday	3

This week is representative of how many hours Ricki will work each week during the summer. If she earns $9 per hour and there are 12 weeks during her summer vacation, what will Ricki's gross earnings be? Explain your reasoning.

9. Table Item Determine whether the quotient of each expression will be positive or negative. **(Lesson 4)**

	positive	negative
$-14 \div (-2)$		
$\frac{64}{-4}$		
$1{,}256 \div 8$		

10. Equation Editor Evaluate $\frac{m}{p}$ if $m = -150$ and $p = -15$. **(Lesson 4)**

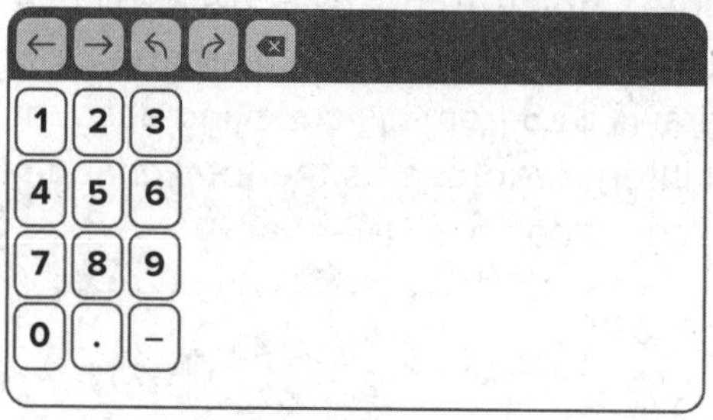

11. Multiple Choice Determine the value of the expression $\frac{-56}{-8} + (-3)$. **(Lesson 5)**

Ⓐ −10

Ⓑ −4

Ⓒ 4

Ⓓ 10

12. Open Response The equation $C = \frac{5(F - 32)}{9}$ can be used to convert temperatures in degrees Fahrenheit to degrees Celsius. **(Lesson 5)**

State of Water	Temperature (°F)
Freezing Point	32
Boiling Point	212

Find the freezing point and boiling point of water in degrees Celsius.

Module 4

Operations with Rational Numbers

Essential Question

How are operations with rational numbers related to operations with integers?

What Will You Learn?

Place a checkmark (✓) in each row that corresponds with how much you already know about each topic **before** starting this module.

KEY

□ — I don't know. ◇ — I've heard of it. ☆ — I know it!

	Before			After		
	□	◇	☆	□	◇	☆
writing fractions as decimals						
adding rational numbers						
subtracting rational numbers						
multiplying rational numbers						
dividing rational numbers						
simplifying expressions involving rational numbers using the order of operations						
evaluating algebraic expressions involving rational numbers						

Foldables Cut out the Foldable and tape it to the Module Review at the end of the module. You can use the Foldable throughout the module as you learn about operations with rational numbers.

What Vocabulary Will You Learn?

Check the box next to each vocabulary term that you may already know.

- ☐ bar notation
- ☐ multiplicative inverses
- ☐ rational number
- ☐ repeating decimal
- ☐ terminating decimal

Are You Ready?

Study the Quick Review to see if you are ready to start this module.
Then complete the Quick Check.

Quick Review

Example 1

Write an equivalent fraction.

Write an equivalent fraction to $\frac{25}{100}$.

$$\frac{25}{100} \overset{\div 25}{=} \frac{1}{4}$$

Divide the numerator and denominator by 25.

An equivalent fraction is $\frac{1}{4}$.

Example 2

Graph mixed numbers on a number line.

Graph $3\frac{2}{3}$ on a number line.

Find the two whole numbers between which $3\frac{2}{3}$ lies.

$3 < 3\frac{2}{3} < 4$

Because the denominator is 3, divide each space into 3 sections.

Draw a dot at $3\frac{2}{3}$.

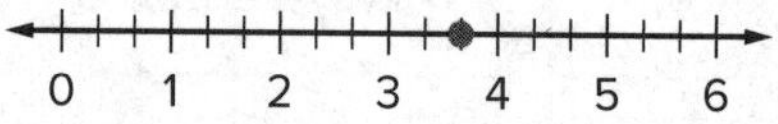

Quick Check

1. Write an equivalent fraction to $\frac{24}{36}$.

2. Write an equivalent fraction to $\frac{45}{50}$.

3. Graph $1\frac{1}{4}$ on a number line.

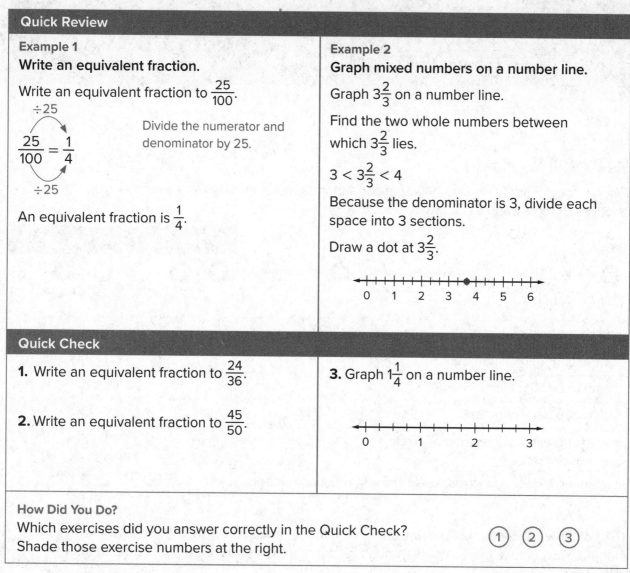

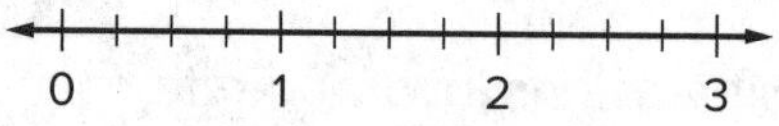

How Did You Do?

Which exercises did you answer correctly in the Quick Check? Shade those exercise numbers at the right.

① ② ③

Rational Numbers

I Can... divide rational numbers and convert fractions to decimal equivalents using division.

What Vocabulary Will You Learn?
bar notation
rational number
repeating decimal
terminating decimal

Learn Rational Numbers

A **rational number** is any number that can be written in the form $\frac{a}{b}$ where a and b are integers, and $b \neq 0$.

The table explains why each number is a rational number.

Number	Explanation
-15	You can write -15 as the ratio $\frac{-15}{1}$.
-0.8	You can write -0.8 as the ratio $\frac{-8}{10}$.
$-\frac{1}{8}$	You can write $-\frac{1}{8}$ as the ratio $\frac{-1}{8}$.
28%	You can write 28% as the ratio $\frac{28}{100}$.

Talk About It!

Is $-\frac{5}{6}$ the same as $\frac{-5}{6}$ or $\frac{5}{-6}$? Justify your response.

Explore Rational Numbers Written as Decimals

Online Activity You will use Web Sketchpad to explore patterns in the decimal form of rational numbers and make a conjecture about the types of numbers that eventually repeat in zeros.

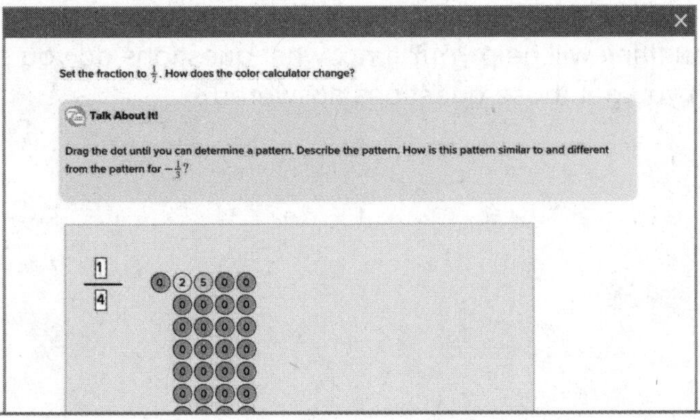

Set the fraction to $\frac{1}{4}$. How does the color calculator change?
Talk About It!
Drag the dot until you can determine a pattern. Describe the pattern. How is this pattern similar to and different from the pattern for $-\frac{1}{3}$?

Your Notes

Math History Minute

Katherine Johnson (1918-) was a research mathematician who was hired to be a "human computer" by the National Advisory Committee for Aeronautics, later known as NASA. While computers had been programmed to control the movements of astronaut John Glenn's space capsule in 1962, Glenn asked Katherine to perform the same calculations by hand to confirm that the computer's calculations were correct.

Learn Rational Numbers Written as Decimals

Any fraction can be expressed as a decimal by dividing the numerator by the denominator. The decimal form of a rational number either terminates in 0s or eventually repeats. **Repeating decimals** are decimals in which 1 or more digits repeat and can be represented using bar notation. In **bar notation**, a bar is drawn only over the digit(s) that repeat.

The following decimals are written in bar notation.

$-0.44444... = -0.\overline{4}$ The digit 4 repeats.

$2.4343... = 2.\overline{43}$ The digits 43 repeat.

Complete the table by writing each decimal using bar notation.

Decimal	Bar Notation
0.11111.....	
0.61111.....	
0.616161.....	
6.160000.....	
6.1611611611.....	

Every decimal can be considered a repeating decimal. Decimals with a repeating digit of zero are also called **terminating decimals**, because the repeating zeros in a terminating decimal are usually truncated, or dropped. For example, the terminating decimal $0.25\overline{0}$ is written as 0.25. The decimal $0.25\overline{0}$ can be considered repeating because the digit 0 repeats.

Pause and Reflect

Are you ready to move on to the Example? If yes, what have you learned that you think will help you? If no, what questions do you still have? How can you get those questions answered?

Record your observations here

Example 1 Write Fractions as Decimals

Write $\frac{1}{40}$ as a decimal. Determine if the decimal is a terminating decimal.

Part A Write the fraction as a decimal.

Divide 1 by 40 using long division.

$40\overline{)1.0000}$

So, $\frac{1}{40} = 0.0250...$ or 0.025.

Part B Determine if the decimal is a terminating decimal.

$\frac{1}{40} = 0.0250...$

The decimal ends with __________ zeros.

So, this is a terminating decimal.

Check

Part A Which of the following is the decimal form of $-\frac{1}{25}$?

Ⓐ -0.04

Ⓑ $-0.0\overline{4}$

Ⓒ $-0.\overline{0}4$

Ⓓ $-0.\overline{04}$

Part B Determine if $-\frac{1}{25}$ is a terminating decimal.

Show your work here

Go Online You can complete an Extra Example online.

Think About It!

Can you predict whether the decimal will be terminating or repeating?

Example 2 Write Fractions as Decimals

Write $-\frac{5}{6}$ as a decimal. Determine if the decimal is a terminating decimal.

Part A Write the fraction as a decimal.

Divide 5 by 6 using long division.

$6\overline{)5.000}$

The remainder of 2 will repeat, so the 3 in the quotient will also repeat.

So, $-\frac{5}{6} = -0.8333....$

Part B Determine if the decimal is a terminating decimal.

The remainder is never zero, so the quotient will have a repeating 3. Because the decimal repeats, write it using bar notation.

$-\frac{5}{6} =$ ______ or ______

So, this is not a terminating decimal.

Check

Write $\frac{5}{9}$ as a decimal. Determine if the decimal is a terminating decimal.

Part A Write $\frac{5}{9}$ as a decimal.

Ⓐ 0.5

Ⓑ 0.59

Ⓒ $0.\overline{5}$

Ⓓ $0.\overline{59}$

Part B Determine if the decimal is a terminating decimal.

Show your work here

Think About It!
How does the denominator of the fraction help you determine if the decimal will terminate?

Talk About It!
If the fraction is negative, how will this affect how it is written in decimal form?

Go Online You can complete an Extra Example online.

Apply Crafting

Two vendors at a craft show are selling signs. The lengths of signs offered by Signs & More and Wood Works are shown in the table. Each sign has the same price. Henry wants to buy the longest of each kind of sign. From which vendor should he buy each style of sign?

	Signs & More	Wood Works
rounded corner sign	$16\frac{1}{8}$ in.	16.25 in.
square corner sign	$16\frac{1}{3}$ in.	16.3 in.

1 What is the task?

Make sure you understand exactly what question to answer or problem to solve. You may want to read the problem three times. Discuss these questions with a partner.

First Time Describe the context of the problem, in your own words.
Second Time What mathematics do you see in the problem?
Third Time What are you wondering about?

2 How can you approach the task? What strategies can you use?

Record your observations here

Talk About It!

Did you convert any of the numbers to a different form in order to solve the problem? Explain.

3 What is your solution?

Use your strategy to solve the problem.

Show your work here

4 How can you show your solution is reasonable?

Write About It! Write an argument that can be used to defend your solution.

Check

During field day, a school had a long jump contest. The top six results of the competition are shown in the table. The teacher wants to award a ribbon to the longest jump by both a 6th grader and a 7th grader, as well as a trophy for the longest jump overall. Who receives the ribbons? Who receives the trophy?

	Grade Level	Length of Jump (ft)
Bentley	7th	$9\frac{5}{8}$
Grant	6th	$9.\overline{68}$
Luna	7th	$9\frac{2}{3}$
Miguel	6th	$9\frac{3}{5}$
Reece	7th	9.5
Trevor	6th	9.73

Show your work here

Go Online You can complete an Extra Example online.

Pause and Reflect

Create a graphic organizer to record the steps to writing a fraction as a decimal and then determining if the decimal is a terminating decimal.

Record your observations here

Name ____________________ Period __________ Date __________

Practice

Go Online You can complete your homework online.

Write each fraction as a decimal. Determine if the decimal is a terminating decimal.
(Examples 1 and 2)

1. $\frac{5}{8}$

2. $-\frac{3}{4}$

3. $\frac{2}{9}$

4. $-\frac{5}{6}$

5. $-\frac{4}{5}$

6. $\frac{23}{50}$

7. $-\frac{9}{22}$

8. $\frac{17}{24}$

9. $-\frac{1}{33}$

10. $-\frac{11}{40}$

11. $\frac{7}{32}$

12. $-\frac{3}{7}$

Test Practice

13. Open Response Ms. Bradley surveyed her class about their favorite fruits. The results are shown in the table.

Fruit	Fraction of the Class
Apples	$\frac{8}{30}$
Kiwi	$\frac{1}{30}$
Peaches	$\frac{12}{30}$
Strawberries	$\frac{9}{30}$

A. Write the fraction of students who prefer strawberries, as a decimal. Determine if the decimal is a terminating decimal.

B. Write the fraction of students who prefer kiwi, as a decimal. Determine if the decimal is a terminating decimal.

Apply

14. Jessica is making matching book bags for her 4 friends. Each book bag needs $1\frac{7}{8}$ yards of fabric. Which of the fabrics shown in the table can Jessica use to make all the book bags for her friends?

Fabric	Amount of Fabric Available (yd)
Moons and Stars	$7\frac{1}{2}$
Softballs	7.4
Stripes	$7\frac{3}{5}$
Tie-Dye	7.9

15. The table shows the times of runners completing a marathon. To qualify for the next marathon, a runner's time must be less than $3\frac{1}{4}$ hours. Which runners qualify?

Runner	Time (h)
Cho	$3\frac{1}{5}$
Kevin	3.2
Ojas	$3\frac{8}{30}$
Sydney	$3\frac{13}{60}$

16. MP **Identify Structure** Write a fraction that is equivalent to a terminating decimal between 0.25 and 0.50.

17. MP **Justify Conclusions** Are there any rational numbers between $0.\overline{5}$ and $\frac{5}{9}$? Justify your answer.

18. MP **Use Math Tools** Eve is making pizza that calls for $\frac{2}{5}$-pound of feta cheese. The store only has packages that contain 0.375- and 0.5- pound of feta cheese. Which of the following strategies might Eve use to determine which package to buy? Use the strategy to solve the problem.

mental math, number sense, estimation

19. MP **Make a Conjecture** Write the following fractions as decimals: $\frac{2}{9}$, $\frac{50}{99}$, and $\frac{98}{99}$. Make a conjecture about how to express these kinds of fractions as decimals.

Add Rational Numbers

I Can... find the additive inverse of a rational number and add rational numbers.

Learn Rational Numbers and Additive Inverses

Two rational numbers are opposites if they are represented on a number line by points that are the same distance but on opposite sides from zero.

Two points, $\frac{3}{4}$ and $-\frac{3}{4}$, are graphed. They are opposites because they are both $\frac{3}{4}$-unit from zero.

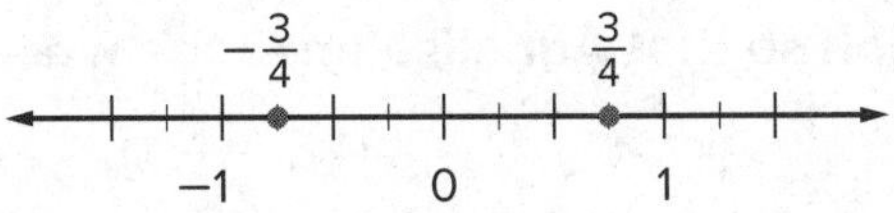

The sum of a number and its opposite, or additive inverse, is zero. The number line shows $-\frac{3}{4} + \frac{3}{4} = 0$.

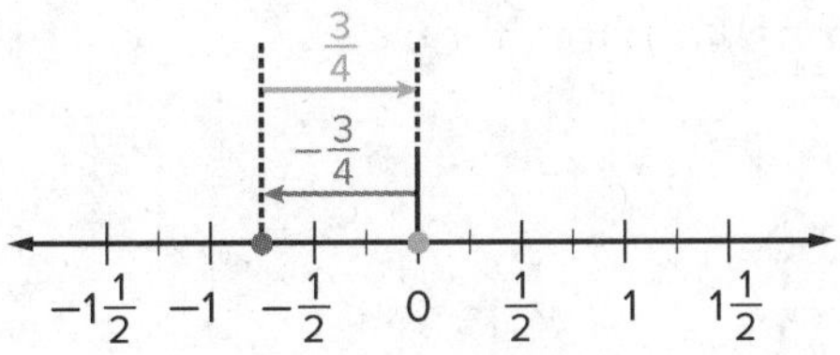

Example 1 Find Additive Inverses

Find the additive inverse of $-\frac{7}{8}$.

Graph and label a point that is the same distance from zero as $-\frac{7}{8}$.

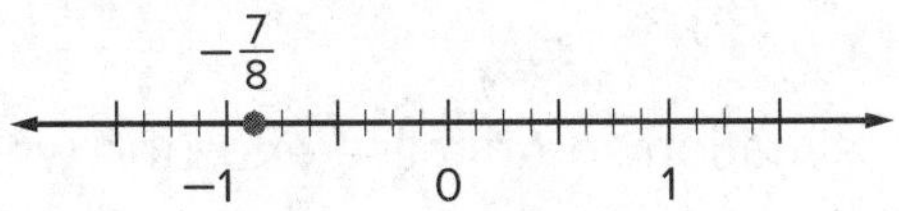

So, the additive inverse of $-\frac{7}{8}$ is ☐.

Your Notes

Check

Find the additive inverse of $-\frac{1}{8}$.

Show your work here

Go Online You can complete an Extra Example online.

Example 2 Find Additive Inverses

Annalise earned \$36.82 at her part time job, and she earned \$18.50 babysitting.

Find the total amount she earned, the additive inverse, and describe a situation so that Annalise ends the week with zero dollars.

Part A Find the total amount she earned.

If p is the amount of money Annalise earned at her part time job, b is the amount of money Annalise earned while babysitting, and t is the total amount of money earned, then

$t = p + b$

$= \$36.82 + \$18.50$

$= \$$ ______

Part B Find the additive inverse.

Annalise ended the week with \$0. What number could you add to \$55.32 that would result in a sum of \$0?

______

Part C Describe a situation so Annalise ends the week with zero dollars.

Circle the situations that represent −\$55.32.

losing \$55.32	getting a gift of \$55.32
finding \$55.32	earning \$55.32
donating \$55.32	spending \$55.32

Think About It!

Should Annalise's babysitting and part-time job money be represented with positive or negative numbers?

Talk About It!

How do you know that losing \$55.32, donating \$55.32, and spending \$55.32 all represent a negative value?

Check

Zoey spent \$12 on a video and \$25.82 on a poster at the music store. Find the total amount she spent, the additive inverse, and describe a situation so that Zoey ends the week with zero dollars.

Part A What was the total amount? ________

Part B What is the additive inverse of the total? ________

Part C Which describes a situation so that Zoey ends the week with zero dollars?

Ⓐ Zoey earned \$37.82 at her lemonade stand.

Ⓑ Zoey spent \$37.82 on dinner and a movie with friends.

Ⓒ Zoey and a friend split the cost of a video game that cost \$37.82.

Ⓓ Zoey got a \$35 gift card for her birthday.

Go Online You can complete an Extra Example online.

Learn Add Rational Numbers

The rules that apply to adding fractions and decimals also apply to rational numbers. The rules for adding integers also apply to positive and negative rational numbers.

Use the chart to see some strategies for how to add rational numbers written in different forms.

Words	Example
Terminating Decimals $\left(-\frac{1}{4}, \frac{3}{8}, \frac{7}{10} \ldots\right)$	
If the fractions are decimals that terminate, use decimals or fractions to add.	$-\frac{1}{5} + 0.8 = -\frac{1}{5} + \frac{4}{5}$ or $-\frac{1}{5} + 0.8 = -0.2 + 0.8$
Non-Terminating Decimals $\left(-\frac{1}{9}, \frac{2}{3}, \frac{11}{15} \ldots\right)$	
If the fractions are decimals that repeat nonzero digits, use fractions to add.	$\frac{1}{3} + (-0.25) = \frac{1}{3} + \left(-\frac{1}{4}\right)$

When a fraction is negative, the sign may be applied to the fraction, the numerator, or the denominator.

$$-\frac{2}{3} = \frac{-2}{3} = \frac{2}{-3}$$

When you are adding two fractions with negative signs, the sign is usually applied to the numerator.

Talk About It!

Why would it be difficult to add −2.5 and $3.1\overline{6}$?

Example 3 Add Rational Numbers

Find $-3\frac{5}{9} + 1\frac{2}{9}$. Write in simplest form.

$-3\frac{5}{9} + 1\frac{2}{9} = -\frac{32}{9} + \frac{11}{9}$	Rewrite the mixed numbers as improper fractions.
$= \frac{-32 + 11}{9}$	Add the numerators. Assign any negative signs to the numerator.
$= \frac{-21}{9}$	Add.
$= \frac{-7}{3}$ or $-2\frac{1}{3}$	Simplify and rename as a mixed number.

So, the sum of $-3\frac{5}{9} + 1\frac{2}{9}$ is ______ or ______.

Think About It!

Because the signs are different, what do you need to do to the mixed numbers before adding them?

Check

Find $-3\frac{1}{6} + \left(-2\frac{5}{6}\right)$.

Show your work here

Go Online You can complete an Extra Example online.

Talk About It!

Why is it important when adding rational numbers to rewrite mixed numbers as improper fractions?

Example 4 Add Rational Numbers

Find $1\frac{3}{4} + \left(-\frac{1}{6}\right)$. Write in simplest form.

$1\frac{3}{4} + \left(-\frac{1}{6}\right) = \square + \left(-\frac{1}{6}\right)$	Rewrite the mixed number as an improper fraction.
$= \frac{21}{12} + \left(-\frac{2}{12}\right)$	The LCD of 4 and 6 is 12.
$= \frac{21 + (-2)}{12}$	Add the numerators. Assign any negative sign to the numerator.
$= \square$ or $\square$	Add and rename as a mixed number.

So, the sum of $1\frac{3}{4} + \left(-\frac{1}{6}\right)$ is $\frac{19}{12}$ or $1\frac{7}{12}$.

Check

Find $-3\frac{1}{3} + 5\frac{1}{4}$.

Show your work here

Example 5 Add Rational Numbers

Find $-\frac{2}{5} + 2.3$.

Because the addends are written in different forms, you can first write them in the same form. The decimal form of $-\frac{2}{5}$ is a terminating decimal. So you can either write the addends as fractions or as decimals.

Method 1 Write both addends as decimals.

$-\frac{2}{5} + 2.3 = \square + 2.3$ Rewrite $-\frac{2}{5}$ as a decimal.

$= \square$ Simplify.

Method 2 Write both addends as fractions.

Go Online Watch the animation to see how to add the two numbers by writing them both as fractions.

$-\frac{2}{5} + 2.3 = -\frac{2}{5} + 2\frac{3}{10}$ Rewrite the decimal as a mixed number.

$= -\frac{4}{10} + \frac{23}{10}$ The LCD of 5 and 10 is 10.

$= \frac{-4 + 23}{10}$ Add the numerators. Assign the negative sign to the numerator.

$= \square$ or $\square$ Add and rename as a mixed number.

So, the sum of $-\frac{2}{5} + 2.3$ is $\frac{19}{10}$ or $1\frac{9}{10}$.

Check

Find $\frac{2}{5} + (-0.75)$.

Show your work here

Go Online You can complete an Extra Example online.

Think About It!

How would you begin evaluating this expression?

Talk About It!

In Method 1, you wrote both addends as decimals before adding. Give an example of when Method 1 is not the best method to use. Explain your reasoning.

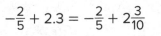

Learn Add Rational Numbers

When adding three or more rational numbers, use the Commutative Property to group numbers by signs or forms.

Grouping numbers with the same form can help simplify the expression.

$\frac{1}{2} + 0.75 + \left(-2\frac{1}{3}\right) + (-3.7)$	Write the expression.
$\frac{1}{2} + 0.75 + \left(-2\frac{1}{3}\right) + (-3.7)$	Identify each number's form as either a fraction or a decimal.
$\frac{1}{2} + \left(-2\frac{1}{3}\right) + 0.75 + (-3.7)$	Rewrite the expression, grouping the numbers by their form.

Talk About It!

How does grouping numbers by like forms help in simplifying an addition expression?

Example 6 Add Rational Numbers

Find $1.25 + \left(-\frac{1}{3}\right) + \left(-\frac{7}{8}\right)$. Write in simplest form.

$1.25 + \left(-\frac{1}{3}\right) + \left(-\frac{7}{8}\right) = \frac{5}{4} + \left(-\frac{1}{3}\right) + \left(-\frac{7}{8}\right)$	Rewrite as an improper fraction.
$= \frac{30}{24} + \left(-\frac{8}{24}\right) + \left(-\frac{21}{24}\right)$	The LCD of 3, 4, and 8 is 24.
$= \frac{30 + (-8) + (-21)}{24}$	Add the numerators.
$= \frac{30 + (-29)}{24}$	Associative Property
$= \frac{1}{24}$	Add.

So, the sum of $1.25 + \left(-\frac{1}{3}\right) + \left(-\frac{7}{8}\right)$ is ______.

Think About It!

Would it be more beneficial to convert the decimals to fractions, or fractions to decimals? Be able to justify your reasoning.

Check

Find $-\frac{1}{6} + 0.20 + 2\frac{5}{9}$.

Show your work here

Go Online You can complete an Extra Example online.

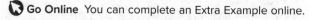

Example 7 Add Rational Numbers

The lowest recorded elevation in the contiguous United States, 282 feet below sea level, is in Death Valley. Suppose you began a hike in the parking area near Badwater Basin at 266 feet below sea level. The stopping points of the hike and the elevation traveled are shown in the table.

What is the elevation of your stopping point?

Stop	Elevation Traveled (ft)
1	−12.5
2	$26\frac{3}{5}$
3	$-3\frac{1}{2}$
4	397.3

Because the denominators of the mixed numbers are 2 and 5, it is easy to convert those to decimals. Write all of the numbers as decimals.

$-266 + (-12.5) + 26\frac{3}{5} + \left(-3\frac{1}{2}\right) + 397.3$ Write the expression.

$= -266 + (-12.5) + 26.6 + (-3.5) + 397.3$ Write fractions as decimals.

$= -266 + (-12.5) + (-3.5) + 26.6 + 397.3$ Commutative Property

$= [-266 + (-12.5) + (-3.5)] + [26.6 + 397.3]$ Associative Property

$= -282 + 423.9$ Simplify.

$= 141.9$ Add.

So, your elevation at the end of the hike is ________ feet above sea level.

Think About It!

How will you represent the starting point of the hike in the expression?

Talk About It!

If the starting point was a rational number, such as $-266\frac{1}{9}$, would it have changed how you found the sum?

Check

During the annual Hot Air Balloon Rally, hot air balloon pilots need to track their altitude as they travel. During a flight that began at 98.1 meters above sea level, one pilot tracked and recorded her altitude every half hour.

Time (hr)	Altitude Change (m)
$\frac{1}{2}$	226.86
1	$-66\frac{4}{8}$
$1\frac{1}{2}$	−15.32
2	$172\frac{3}{4}$

What is her altitude, in meters, after two hours?

Show your work here

Go Online You can complete an Extra Example online.

Pause and Reflect

Think about scenarios, when adding rational numbers, where it would be beneficial to change all the numbers to decimals, or all to fractions. Give examples.

Record your observations here

Apply Animal Care

A veterinarian measures the changes in a cat's weight over four months. If the cat weighed 17.25 pounds at its first visit, what is the cat's weight after its last visit?

Month	Change from Previous Month (lb)
1	-0.5
2	$-2\frac{1}{5}$
3	$-\frac{3}{10}$
4	0.35

1 What is the task?

Make sure you understand exactly what question to answer or problem to solve. You may want to read the problem three times. Discuss these questions with a partner.

First Time Describe the context of the problem, in your own words.
Second Time What mathematics do you see in the problem?
Third Time What are you wondering about?

2 How can you approach the task? What strategies can you use?

Record your observations here

3 What is your solution?

Use your strategy to solve the problem.

Show your work here

4 How can you show your solution is reasonable?

Write About It! Write an argument that can be used to defend your solution.

Talk About It!
Why is it more efficient in this problem to write each of the values as a decimal before adding?

Check

A local forest that covers $472\frac{7}{10}$ acres has gone through periods of growth and loss over the past four years. In some years, firefighters started a controlled burn that burns part of the forest in order to encourage new growth. Each year they track the changes in the area of the forest. Find the area of the forest after the four years.

Year	Change in Area (acres)
1	$3\frac{4}{5}$
2	$-1\frac{1}{2}$
3	-7.9
4	$2\frac{2}{5}$

Show your work here

Go Online You can complete an Extra Example online.

Foldables It's time to update your Foldable, located in the Module Review, based on what you learned in this lesson. If you haven't already assembled your Foldable, you can find the instructions on page FL1.

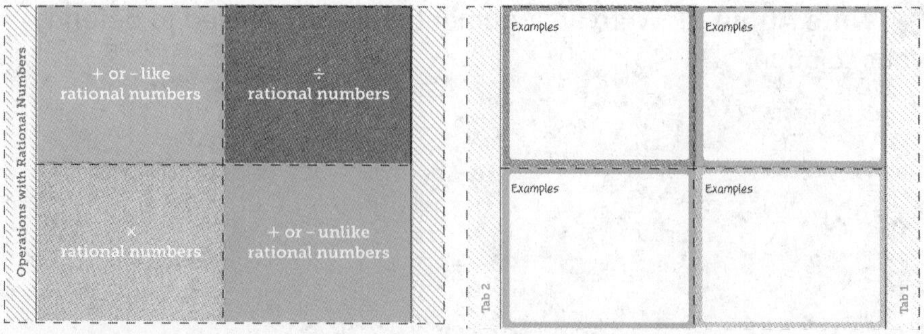

Name ______________________ Period ____________ Date ______________

Practice

Go Online You can complete your homework online.

Find the additive inverse of each rational number. (Example 1)

1. $-\frac{1}{2}$

2. 0.25

3. $\frac{9}{10}$

4. -0.4

5. Quinn earned \$24.50 dog-sitting and \$12.70 for recycling cans. Find the total amount he earned and describe a situation in which Quinn ends the week with zero dollars. **(Example 2)**

6. Rachel spent \$3.89 on a magazine and \$4.86 on a smoothie. Find the total amount she spent and describe a situation in which Rachel ends the day with zero dollars. **(Example 2)**

Add. Write in simplest form. (Examples 3–6)

7. $3\frac{5}{6} + \left(-1\frac{1}{6}\right)$

8. $-13\frac{1}{4} + 4\frac{3}{4}$

9. $-\frac{2}{3} + 2\frac{3}{8}$

10. $2\frac{1}{2} + \left(-\frac{1}{3}\right)$

11. $-3.7 + \frac{1}{4}$

12. $\frac{1}{3} + 4.1$

13. $-1\frac{1}{4} + 0.75 + 0.45$

14. $-0.25 + 3\frac{1}{6} + 2\frac{1}{12}$

15. Marlee is making jewelry for a class craft show. She began with 115 inches of wire. She used 25.75 inches for rings. Then her teacher gave her $30\frac{1}{4}$ inches of wire to make more jewelry. She then used $38\frac{1}{2}$ inches for the bracelets and 60.2 inches for necklaces. How much wire does Marlee have left? **(Example 7)**

Test Practice

16. Open Response The table shows the change in the amount of food for a male sea otter at a zoo. The zookeeper starts with 100.75 pounds of food and tracks how much the otter eats and any food deliveries. What is the food supply after Day 4?

Day	1	2	3	4
Change in Food Supply (pounds)	$-20\frac{1}{16}$	$-23\frac{3}{8}$	-25.25	$75\frac{3}{4}$

Apply

17. Jada measures the changes in her dog's weight over the entire year. She weighs her dog every 3 months and records the results. If Jada's dog weighed 55.75 pounds at the beginning of the year, what is the dog's weight at the end of the year?

Months	Difference from Previous Weight (lb)
January–March	+2.125
April–June	$-3\frac{1}{4}$
July–September	$-\frac{1}{2}$
October–December	+0.875

18. A local petting zoo allows visitors to feed the goats food pellets. The petting zoo starts the month with 525.25 pounds of food pellets. Each week, the workers track the amount of food pellets given out and any food pellet deliveries. Use the table to determine the number of pounds of food pellets the petting zoo will have at the end of the month.

Week	Change in Food Pellets (lb)
1	$-164\frac{1}{2}$
2	−189.75
3	$355\frac{7}{8}$
4	$-200\frac{3}{16}$

19. Write an addition problem with unlike mixed numbers and a least common denominator of 16. Find the sum in simplest form.

20. MP **Justify Conclusions** Is the additive inverse of a number *always, sometimes,* or *never* negative? Justify your answer with an example.

21. MP **Find the Error** A student is adding $1\frac{2}{9}$, $3\frac{1}{3}$, and $-4\frac{5}{6}$. The first step the student performs is to find the common denominator of 9, 3, and 6. Find the student's mistake and correct it.

The least common denominator of 9, 3, and 6 is 36 because you can divide 36 by all of these numbers without getting a remainder.

22. Suppose you use 30 instead of 15 as a common denominator when finding $6\frac{3}{5} + \left(-3\frac{1}{3}\right)$. How will that change the process for finding the sum?

Subtract Rational Numbers

I Can... subtract rational numbers by adding the additive inverse.

Learn Subtract Rational Numbers

To subtract rational numbers in different forms, write the numbers in the same form.

Go Online Watch the animation to see how to subtract a mixed number and a decimal.

The animation shows how to rewrite a problem using decimals. This method works if the mixed number can be rewritten as a terminating decimal.

Rewrite Using Decimals	
Steps	**Example**
1. Write the mixed number or fraction as a decimal.	$2\frac{2}{5} - 6.55 = 2\frac{4}{10} - 6.55$ $= 2.4 - 6.55$
2. Subtract the decimals.	$= 2.4 + (-6.55)$ $= -4.15$

The animation also shows how to rewrite a problem using fractions or mixed numbers. This method works when the fraction or mixed number cannot be written as a terminating decimal.

Rewrite Using Fractions or Mixed Numbers	
Steps	**Example**
1. Write the decimal as a fraction or mixed number.	$4.6 - 2\frac{2}{3} = 4\frac{6}{10} - 2\frac{2}{3}$ $= 4\frac{3}{5} - 2\frac{2}{3}$
2. Rewrite fractions with a common denominator.	$= 4\frac{9}{15} - 2\frac{10}{15}$
3. Subtract the fractions or mixed numbers.	$= 3\frac{24}{15} + \left(-2\frac{10}{15}\right)$
4. Simplify if necessary.	$= 1\frac{14}{15}$

(continued on next page)

Your Notes

Talk About It!

How does knowing how to add rational numbers help you to subtract rational numbers?

Use the chart below to find out how to subtract rational numbers written in different forms.

Words	Example
Terminating Decimals $\left(-\frac{1}{4}, \frac{3}{8}, \frac{7}{8} \ldots\right)$	
If the fractions are decimals that terminate, use decimals or fractions to subtract.	$-0.90 - \frac{1}{10} = -\frac{9}{10} - \frac{1}{10}$ or $= -0.9 - 0.1$
Non-Terminating Decimals $\left(-\frac{1}{9}, \frac{2}{3}, \frac{11}{15} \ldots\right)$	
If the fractions are decimals that repeat nonzero digits, use fractions to subtract.	$-\frac{1}{6} - 0.125 = -\frac{1}{6} - \frac{1}{8}$

Example 1 Subtract Rational Numbers

Find −3.27 − (−6.7).

Use the same rules to subtract positive and negative decimals as subtracting integers.

Integers		Rational Numbers
$-3 - (-6)$	Write the expression.	$-3.27 - (-6.7)$
$-3 + 6$	Add the additive inverse.	$-3.27 + 6.7$
3	Add.	3.43

So, because $|6.7| > |-3.27|$, the sum will have the same sign as 6.7, positive.

So, −3.27 − (−6.7) is ________.

Check

Find −4.2 − 3.57.

Show your work here

Go Online You can complete an Extra Example online.

Example 2 Subtract Rational Numbers

Find $5\frac{1}{3} - \left(-4\frac{5}{9}\right)$. Write in simplest form.

$5\frac{1}{3} - \left(-4\frac{5}{9}\right) = \frac{16}{3} - \left(-\frac{41}{9}\right)$ Write the mixed numbers as improper fractions.

$= \square - \left(\square\right)$ The LCD of 3 and 9 is 9.

$= \frac{48}{9} + \frac{41}{9}$ Add using the additive inverse.

$= \square$ or $\square$ Simplify.

So, the difference of $5\frac{1}{3} - \left(-4\frac{5}{9}\right)$ is $\frac{89}{9}$ or $9\frac{8}{9}$.

Check

Find $3\frac{1}{2} - \left(-1\frac{3}{10}\right)$. Write in simplest form.

Show your work here

Go Online You can complete an Extra Example online.

Pause and Reflect

How can you use inverse operations to check your work?

Record your observations here

Example 3 Evaluate Expressions

Evaluate $x - y$ if $x = -2\frac{4}{5}$ and $y = 1.4$.

Determine if the mixed number is a decimal that terminates. Because $-2\frac{4}{5}$ terminates, you can use either decimals or fractions to subtract.

Method 1 Evaluate using decimals.

$x - y = -2\frac{4}{5} - 1.4$	Replace x with $-2\frac{4}{5}$ and y with 1.4.
$= -2.8 - 1.4$	Rewrite $-2\frac{4}{5}$ as a decimal.
$= -2.8 + (-1.4)$	Add the additive inverse of 1.4.
$= -4.2$	Simplify.

Method 2 Evaluate using fractions.

$x - y = -2\frac{4}{5} - 1.4$	Replace x with $-2\frac{4}{5}$ and y with 1.4.
$= -2\frac{4}{5} - 1\frac{2}{5}$	Rewrite 1.4 as a mixed number.
$= -\frac{14}{5} - \frac{7}{5}$	Write mixed numbers as improper fractions.
$= -\frac{14}{5} + \left(-\frac{7}{5}\right)$	Add the additive inverse of $\frac{7}{5}$.
$= -\frac{21}{5}$ or $-4\frac{1}{5}$	Add.

So, when $x = -2\frac{4}{5}$ and $y = 1.4$, $x - y =$ ______ or ______.

Think About It!

How would you begin to evaluate the expression?

Talk About It!

Generate two different expressions that involve the subtraction of rational numbers. The first expression should be best simplified by converting any decimals to fractions. The second expression should be best simplified by converting fractions to decimals.

Check

Evaluate $x - y$ if $x = 3\frac{3}{4}$ and $y = -4.2$. Write in simplest form.

Show your work here

Go Online You can complete an Extra Example online.

Foldables It's time to update your Foldable, located in the Module Review, based on what you learned in this lesson. If you haven't already assembled your Foldable, you can find the instructions on page FL1.

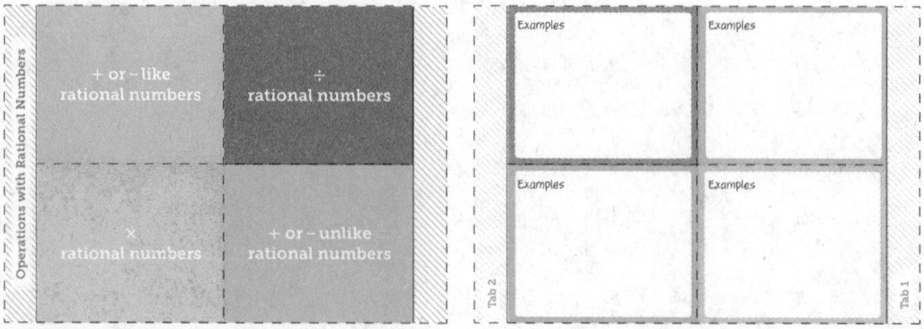

Name ______________________ Period ____________ Date ____________

Practice

Go Online You can complete your homework online.

Subtract. Write in simplest form. **(Examples 1 and 2)**

1. $-2.45 - (-3.9)$

2. $-4.6 - (-2.31)$

3. $5.47 - (-2.8)$

4. $-6.2 - 3.79$

5. $7\frac{5}{12} - \left(-3\frac{3}{4}\right)$

6. $5\frac{9}{10} - \left(-8\frac{2}{5}\right)$

7. $-\frac{7}{8} - 2\frac{1}{6}$

8. $-\frac{8}{15} - 3\frac{4}{5}$

9. $-9\frac{7}{10} - \left(-4\frac{3}{5}\right)$

10. $\frac{5}{6} - \left(-\frac{3}{4}\right)$

11. $-\frac{2}{3} - \left(-\frac{1}{2}\right)$

12. $-\frac{7}{10} - \left(-\frac{4}{15}\right)$

13. Evaluate $x - y$ if $x = -5\frac{1}{4}$ and $y = -6.8$. Write your answer in simplest form. **(Example 3)**

15. Evaluate $s - t$ if $s = \frac{1}{4}$ and $t = -1.75$. Write your answer in simplest form. **(Example 3)**

Test Practice

14. Equation Editor Evaluate $a - b$ if $a = -2.5$ and $b = \frac{2}{5}$. Write your answer in simplest form.

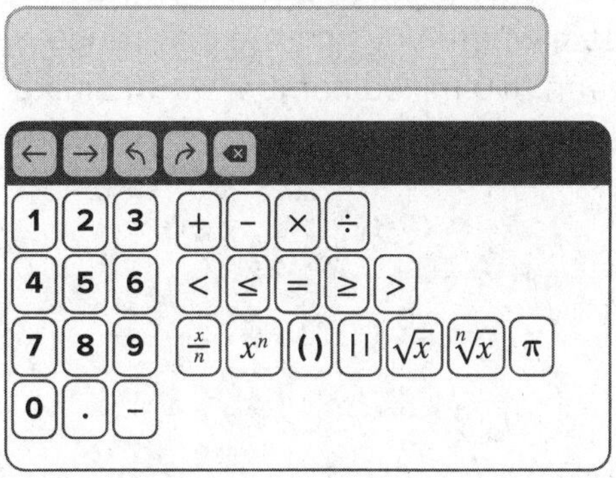

Apply

16. The table shows the allowance balance for the children in the Martinez family. What is the difference between Maria's allowance balance and Charles' allowance balance?

Child	Allowance Balance ($)
Annemarie	−12.50
Charles	−5.75
Maria	−15.25
Sylvester	−8.00

17. The table shows the location below sea level of hikers on a trail. What is the difference between Tommy's location and Theresa's location? Write your answer in simplest form.

Hiker	Location (ft)
Camila	−22.5
Malik	−34
Theresa	$-15\frac{1}{2}$
Tommy	$-20\frac{3}{4}$

18. Write a subtraction problem with unlike mixed numbers and a least common denominator of 10. Find the difference in simplest fraction form.

19. MP **Use a Counterexample** Is the following statement *true* or *false*? If false, provide a counterexample.

The difference between a positive mixed number and negative mixed number is never positive.

20. Create Write and solve a real-world problem where you find the difference between two mixed numbers with unlike denominators.

21. Suppose you use 8 instead of 4 as a common denominator when finding $7\frac{1}{2} - \left(-3\frac{1}{4}\right)$. How will that change the process for finding the difference?

Multiply Rational Numbers

I Can... use the rules for multiplying integers to multiply rational numbers.

Learn Multiply Rational Numbers

The rules for integers also apply to positive and negative rational numbers.

State whether the product of each expression will be *positive* or *negative* using your knowledge of multiplying integers.

Multiply Rational Numbers	
Expression	Sign of Product
$\frac{1}{7}\left(-\frac{2}{3}\right)$	
$2\frac{3}{5} \cdot \frac{5}{7}$	
$-\frac{1}{3} \cdot \frac{7}{9}$	
$-1\frac{1}{2}\left(-5\frac{1}{4}\right)$	

Example 1 Multiply Rational Numbers

Find $-\frac{3}{4}\left(-\frac{7}{9}\right)$. Write in simplest form.

$= -\frac{\overset{1}{\cancel{3}}}{4}\left(-\frac{7}{\underset{3}{\cancel{9}}}\right)$ Divide by common factors.

$= \frac{-1(-7)}{4 \cdot 3}$ Multiply the numerators and denominators.

$=$ ☐ Simplify.

So, $-\frac{3}{4}\left(-\frac{7}{9}\right)$ is $\frac{7}{12}$.

Your Notes

Check

Find $-\frac{7}{8}\left(\frac{4}{14}\right)$.

Show your work here

Go Online You can complete an Extra Example online.

Think About It!

Before you begin to multiply the rational numbers, what do you need to do?

Example 2 Multiply Rational Numbers

Find $-3\frac{1}{5}\left(1\frac{1}{14}\right)$. Write in simplest form.

$= -\frac{16}{5}\left(\frac{15}{14}\right)$ Write the mixed numbers as improper fractions.

$= -\frac{^{8}\cancel{16}}{\cancel{5}_{1}}\left(\frac{\cancel{15}^{3}}{\cancel{14}_{7}}\right)$ Divide by common factors.

$= \frac{-8 \cdot 3}{1 \cdot 7}$ Multiply the numerators and denominators.

$=$ ☐ or ☐ Simplify.

So, the product of $-3\frac{1}{5}\left(1\frac{1}{14}\right)$ is $-\frac{24}{7}$ or $-3\frac{3}{7}$.

Talk About It!

How would the steps be different if you did not simplify the rational numbers before multiplying?

Check

Find $-1\frac{7}{8}\left(-2\frac{2}{5}\right)$. Write in simplest form.

Show your work here

Go Online You can complete an Extra Example online.

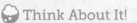

Learn Multiply Rational Numbers

When multiplying rational numbers written in different forms, write the factors in the same form.

Words	Example
Terminating Decimals $\left(-\frac{1}{4}, \frac{3}{8}, \frac{7}{8}\ldots\right)$	
If the fractions or mixed numbers are decimals that terminate, use decimals or fractions to multiply.	$-3.7\left(1\frac{1}{5}\right) = -3.7 \cdot (1.2)$ or $= -3\frac{7}{10}\left(1\frac{1}{5}\right)$
Non-Terminating Decimals $\left(-\frac{1}{9}, \frac{2}{3}, \frac{11}{15}\ldots\right)$	
If the fractions or mixed numbers are decimals that repeat nonzero digits, use fractions to multiply.	$0.75\left(-\frac{2}{3}\right) = \frac{3}{4}\left(-\frac{2}{3}\right)$

Pause and Reflect

Create a graphic organizer that will help you understand how to determine if a fraction or mixed number is a terminating or non-terminating decimal.

Record your observations here

Example 3 Multiply Rational Numbers

Find $\frac{1}{3}(-2.75)$. Write in simplest form.

Because the factors are written in different forms, you first need to rewrite them in the same form.

Because $\frac{1}{3}$ repeats non-zero digits, write the second factor as a mixed number.

$\frac{1}{3}(-2.75)$	Write the expression.
$= \frac{1}{3}\left(-2\frac{3}{4}\right)$	Write -2.75 as a mixed number.
$= \frac{1}{3}\left(-\frac{11}{4}\right)$	Write $-2\frac{3}{4}$ as an improper fraction.
$= \frac{1(-11)}{(3 \cdot 4)}$	Multiply the numerators and denominators.
$= \square$	Simplify.

So, the product of $\frac{1}{3}(-2.75)$ is $-\frac{11}{12}$.

Check

Find $-0.64\left(-\frac{1}{4}\right)$.

Show your work here

Go Online You can complete an Extra Example online.

Pause and Reflect

How do you know, without calculating, that $\frac{1}{3}$ repeats non-zero digits?

Record your observations here

Think About It!

Will you write both numbers as fractions or as decimals?

Talk About It!

Is it possible to find the product without writing the numbers in the same form? Explain.

Example 4 Multiply Rational Numbers

Evaluate $\frac{1}{2}ab$ if $a = 1\frac{3}{7}$ and $b = -\frac{4}{9}$.

$\frac{1}{2}ab = \frac{1}{2} \cdot 1\frac{3}{7}\left(-\frac{4}{9}\right)$ Replace a with $1\frac{3}{7}$ and b with $-\frac{4}{9}$.

$= \frac{1}{2} \cdot \frac{10}{7}\left(-\frac{4}{9}\right)$ Write $1\frac{3}{7}$ as an improper fraction.

$= \frac{1}{\not{2}_1} \cdot \frac{\not{10}^5}{7}\left(-\frac{4}{9}\right)$ Divide by common factors.

$= \frac{1 \cdot 5\,(-4)}{1 \cdot 7 \cdot 9}$ Multiply the numerators and denominators.

$=$ ☐ Simplify.

So, the value of $\frac{1}{2}ab$ when $a = 1\frac{3}{7}$ and $b = -\frac{4}{9}$ is $-\frac{20}{63}$.

Check

Evaluate $\frac{2}{3}xy$ if $x = -3\frac{1}{6}$ and $y = -5\frac{1}{4}$.

Show your work here

Go Online You can complete an Extra Example online.

Pause and Reflect

When multiplying rational numbers, why is it beneficial to divide by common factors?

Record your observations here

Example 5 Multiply Rational Numbers

Evaluate $\frac{1}{4}xyz$ if $x = 9.5$, $y = -0.8$, and $z = 2\frac{1}{5}$.

Determine if the fraction and mixed number are terminating decimals. Because $\frac{1}{4}$ and $2\frac{1}{5}$ both terminate, you can use either decimals or fractions to multiply.

$\frac{1}{4}xyz = \frac{1}{4}(\square)(\square)(\square)$ — Replace x with 9.5, y with -0.8, and z with $2\frac{1}{5}$.

$= 0.25(9.5)(-0.8)(2.2)$ — Replace $\frac{1}{4}$ with 0.25 and $2\frac{1}{5}$ with 2.2.

$= \square$ — Simplify.

So, the value of $\frac{1}{4}xyz$ when $x = 9.5$, $y = -0.8$, and $z = 2\frac{1}{5}$ is -4.18.

Think About It!

What is the sign of the product of the four numbers?

Talk About It!

How would your answer compare if you had multiplied using fractions rather than decimals?

Check

Evaluate abc if $a = -2.8$, $b = -2\frac{1}{7}$, and $c = -\frac{3}{13}$.

Show your work here

Go Online You can complete an Extra Example online.

Pause and Reflect

How can you use estimation to check your work?

Record your observations here

Apply Temperature

A sign shows the temperature in Badger, Minnesota, at 10 P.M. is −11.5°F. At 4 A.M., the temperature changed by $\frac{2}{3}$ of the current value. What is the final temperature?

1 What is the task?

Make sure you understand exactly what question to answer or problem to solve. You may want to read the problem three times. Discuss these questions with a partner.

First Time Describe the context of the problem, in your own words.
Second Time What mathematics do you see in the problem?
Third Time What are you wondering about?

2 How can you approach the task? What strategies can you use?

Record your observations here

3 What is your solution?

Use your strategy to solve the problem.

Show your work here

4 How can you show your solution is reasonable?

Write About It! Write an argument that can be used to defend your solution.

Talk About It!
Is the final temperature warmer or colder than the original temperature? Explain.

Check

Lena was practicing freediving. On her first attempt, she dove to a depth of −63.4 feet. Her second attempt changed by $\frac{2}{5}$ of the original depth. How far did Lena dive on her second attempt?

Show your work here

Go Online You can complete an Extra Example online.

Foldables It's time to update your Foldable, located in the Module Review, based on what you learned in this lesson. If you haven't already assembled your Foldable, you can find the instructions on page FL1.

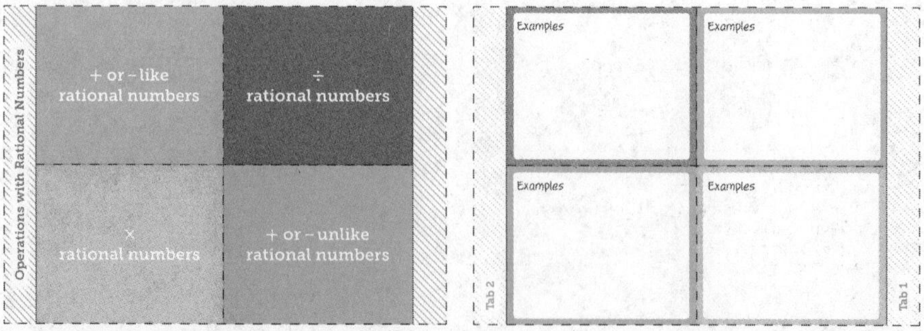

Name ____________________ Period __________ Date __________

Practice

Go Online You can complete your homework online.

Multiply. Write the product in simplest form. **(Examples 1 and 2)**

1. $-\frac{1}{2}\left(-\frac{4}{5}\right)$

2. $-\frac{3}{8}\left(-\frac{8}{9}\right)$

3. $-\frac{1}{4}\left(-\frac{4}{5}\right)$

4. $1\frac{4}{9}\left(-2\frac{4}{7}\right)$

5. $1\frac{1}{10}\left(-6\frac{7}{8}\right)$

6. $-5\frac{1}{4}\left(-4\frac{2}{3}\right)$

Multiply. Write the product in simplest form. **(Example 3)**

7. $-\frac{1}{6}(2.4)$

8. $\frac{2}{5}(-3.75)$

9. $-\frac{1}{4}(-8.6)$

Evaluate each expression if $x = -\frac{2}{3}$, $y = \frac{3}{5}$, and $z = -1\frac{7}{8}$. Write the product in simplest form. **(Example 4)**

10. $\frac{1}{4}xy$

11. $-\frac{4}{5}xz$

12. $\frac{1}{2}yz$

13. Evaluate $-xyz$ if $x = -8.4$, $y = 0.25$, and $z = 3\frac{4}{5}$. Write your answer in simplest form. **(Example 5)**

Test Practice

14. Equation Editor Evaluate $\frac{1}{2}xyz$ if $x = -8.4$, $y = 0.25$, and $z = 3\frac{4}{5}$. Write your answer in simplest form.

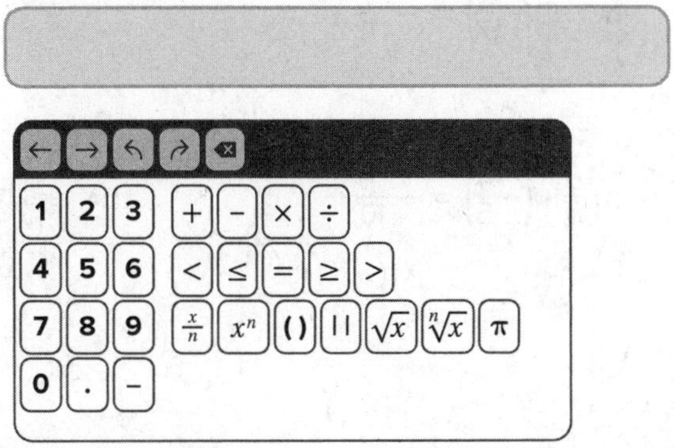

Apply

15. The table shows the change in the value of Rudo's stocks one day. The next day, the value of the All-Plus stock dropped $\frac{1}{4}$ of the amount it changed from the previous day. What was the total change in the All-Plus stock? Round to the nearest cent.

Stock	Change
All-Plus	−\$2.50
True-Fit	−\$3.75

16. The table shows the temperature at midnight in Fink Creek, Alaska. At 5 A.M., the temperature changed by $\frac{1}{2}$ of the current value. What is the temperature at 5 A.M.?

Time	Temperature (°F)
12:00 A.M	−13.4
5:00 A.M	?

17. MP **Reason Inductively** Find two rational numbers greater than $\frac{1}{4}$ whose product is less than $\frac{1}{4}$. Explain.

18. MP **Use a Counterexample** Is the following statement *true* or *false*? If false, provide a counterexample.

The product of a fraction between 0 and 1 and a whole number or mixed number is never less than the whole number or mixed number.

19. MP **Persevere with Problems** Find the missing fraction for each problem.

a. $-\frac{6}{11} \cdot \left(\frac{x}{y}\right) = -\frac{1}{11}$

b. $\frac{a}{b} \cdot \left(-\frac{4}{5}\right) = -\frac{1}{10}$

20. Estimate the product of 2.6 and $3\frac{7}{10}$. Then find the actual product. Are the estimate and the actual product different? If so, how could you improve the estimate?

Divide Rational Numbers

I Can... use the rules for dividing integers to divide rational numbers.

What Vocabulary Will You Learn?
multiplicative inverses

Learn Divide Rational Numbers

The rules for integers also apply to positive and negative rational numbers.

Two numbers whose product is 1 are called **multiplicative inverses**, or reciprocals. For example, $-\frac{1}{4}$ and -4 are multiplicative inverses because $-\frac{1}{4}(-4) = 1$.

Go Online Watch the animation to see how to divide rational numbers.

Dividing Rational Numbers	
Steps	**Example**
1. Rewrite the division as multiplication by the reciprocal.	$3 \div \left(-\frac{1}{4}\right) = -12$ reciprocals ↕ ↕ same result
2. Multiply.	$3 \cdot (-4) = -12$
3. Identify the quotient.	

Talk About It!
Why is the reciprocal of a fraction with 1 in the numerator an integer?

Example 1 Divide Rational Numbers

Find $-\frac{2}{3} \div \frac{1}{9}$.

$-\frac{2}{3} \div \frac{1}{9} = -\frac{2}{3} \cdot \frac{9}{1}$ Multiply by the multiplicative inverse.

$= -\frac{2}{\cancel{3}_1}\left(\frac{\cancel{9}^3}{1}\right)$ Divide by common factors.

$= -\frac{6}{1}$ or -6 Simplify.

So, the solution to $-\frac{2}{3} \div \frac{1}{9}$ is ________.

Your Notes

Check

Find $-\frac{6}{7} \div 12$.

Show your work here

Go Online You can complete an Extra Example online.

Think About It!

Before you divide the two rational numbers, what do you need to do?

Example 2 Divide Rational Numbers

Find $-4\frac{2}{5} \div \left(-2\frac{14}{15}\right)$.

$-4\frac{2}{5} \div \left(-2\frac{14}{15}\right) = -\frac{22}{5} \div \left(-\frac{44}{15}\right)$ Write the mixed numbers as improper fractions.

$= -\frac{22}{5}\left(-\frac{15}{44}\right)$ Multiply by the multiplicative inverse.

$= -\frac{\overset{1}{\cancel{22}}}{\underset{1}{\cancel{5}}}\left(-\frac{\overset{3}{\cancel{15}}}{\underset{2}{\cancel{44}}}\right)$ Divide by common factors.

$= \frac{3}{2}$ or $1\frac{1}{2}$ Simplify.

Talk About It!

How do you know the quotient will be a positive number before you divide?

So, $-4\frac{2}{5} \div \left(-2\frac{14}{15}\right)$ is ☐ or ☐.

Check

Find $-1\frac{1}{3} \div 2\frac{2}{15}$.

Show your work here

Go Online You can complete an Extra Example online.

Learn Divide Rational Numbers

When dividing rational numbers written in different forms, write the factors in the same form.

Words	Example
Terminating Decimals $\left(-\frac{1}{4}, \frac{3}{8}, \frac{7}{8}...\right)$	
If the fractions or mixed numbers are decimals that terminate, use decimals or fractions to divide.	$6.3 \div \left(-\frac{7}{8}\right) = 6.3 \div -0.875$ or $6.3 \div \left(-\frac{7}{8}\right) = 6\frac{3}{10} \div \left(-\frac{7}{8}\right)$
Non-Terminating Decimals $\left(-\frac{1}{9}, \frac{2}{3}, \frac{11}{15}...\right)$	
If the fractions or mixed numbers are decimals that repeat nonzero digits, use fractions to divide.	$0.75 \div \left(-\frac{2}{3}\right) = \frac{3}{4} \div \left(-\frac{2}{3}\right)$

Pause and Reflect

Are you ready to move on to the next Example? If yes, what have you learned that you think will help you? If no, what questions do you still have? How can you get those questions answered?

Record your observations here

Think About It!

How will you decide whether to evaluate the expression with decimals or fractions?

Talk About It!

Is it more advantageous to use one method over the other? Explain.

Example 3 Divide Rational Numbers

Evaluate $\frac{a}{b}$ if $a = \frac{3}{4}$ and $b = -0.05$.

Method 1 Evaluate with numbers as decimals.

$\frac{a}{b} = \frac{\frac{3}{4}}{-0.05}$ Replace a with $\frac{3}{4}$ and b with -0.05.

$= \frac{\square}{-0.05}$ Write $\frac{3}{4}$ as a decimal.

$= -15$ Divide.

Method 2 Evaluate with numbers as fractions.

$\frac{a}{b} = \frac{\frac{3}{4}}{-0.05}$ Replace a with $\frac{3}{4}$ and b with -0.05.

$= \frac{\frac{3}{4}}{\left(-\frac{1}{20}\right)}$ Write -0.05 as a fraction.

$= \frac{3}{4} \div \left(\square\right)$ Write the complex fraction as a division problem.

$= \frac{3}{4}\left(\square\right)$ Multiply by the multiplicative inverse.

$= \frac{3}{\cancel{4}_1}\left(-\frac{\cancel{20}^5}{1}\right)$ Divide by common factors.

$= \frac{3(-5)}{1 \cdot 1}$ Multiply the numerators and denominators.

$= \square$ Simplify.

So, the solution is -15.

Check

Evaluate $\frac{x}{y}$ if $x = -6.85$ and $y = -2\frac{2}{5}$. If your answer is in decimal form, round to the nearest hundredth.

Show your work here

Go Online You can complete an Extra Example online.

Apply Finance

A stock market simulation project was assigned in Susannah's math class. She was given $1,000 to invest. She invested her money into a retail electronics business and over the course of 6 months she tracked her gains and losses. If Susannah owned 100 shares of the company, what would the average change of the value of her shares be, in dollars?

Month	Gain/Loss per Share
January	−$5.36
February	$1.20
March	$1.85
April	−$0.63
May	$2.16
June	−$3.72

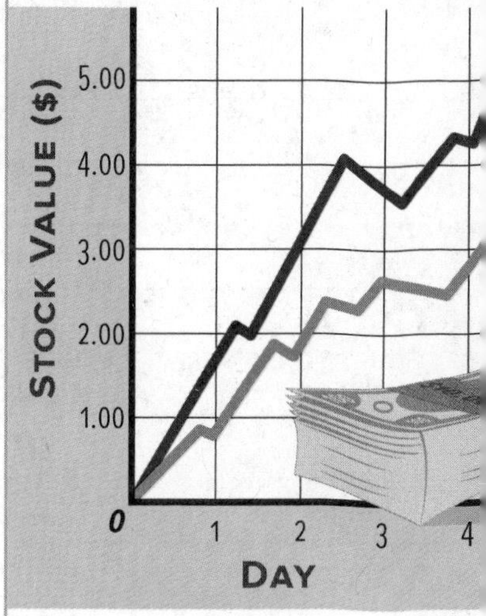

1 What is the task?

Make sure you understand exactly what question to answer or problem to solve. You may want to read the problem three times. Discuss these questions with a partner.

First Time Describe the context of the problem, in your own words.
Second Time What mathematics do you see in the problem?
Third Time What are you wondering about?

2 How can you approach the task? What strategies can you use?

Record your observations here

3 What is your solution?

Use your strategy to solve the problem.

Show your work here

4 How can you show your solution is reasonable?

Write About It! Write an argument that can be used to defend your solution.

Talk About It!

How do you find the average change of the value of her shares?

Check

The student council is selling lollipops to fund the upcoming school dance. The table shows their results over five weeks.

Week	Profit ($)
1	−13.42
2	25.90
3	39.53
4	−17.52
5	42.31

What was the average amount of money the student council earned each week?

Show your work here

Go Online You can complete an Extra Example online.

Foldables It's time to update your Foldable, located in the Module Review, based on what you learned in this lesson. If you haven't already assembled your Foldable, you can find the instructions on page FL1.

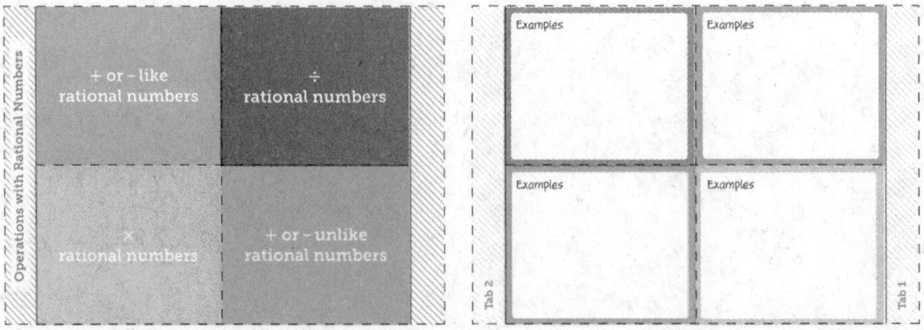

Name ______________________ Period ____________ Date ____________

Practice

Go Online You can complete your homework online.

Divide. Write your answer in simplest fraction form. **(Examples 1 and 2)**

1. $-\frac{6}{7} \div \frac{3}{14}$

2. $-\frac{3}{4} \div \left(-\frac{1}{2}\right)$

3. $\frac{2}{3} \div \left(-\frac{4}{9}\right)$

4. $-\frac{4}{7} \div \frac{8}{9}$

5. $-\frac{3}{5} \div \left(-\frac{3}{4}\right)$

6. $-5\frac{3}{4} \div \left(2\frac{1}{16}\right)$

7. $7\frac{1}{2} \div \left(-2\frac{5}{6}\right)$

8. $-3\frac{4}{9} \div \left(-2\frac{1}{3}\right)$

9. $2\frac{2}{3} \div \left(-1\frac{1}{6}\right)$

10. $-3\frac{2}{5} \div \left(-5\frac{1}{10}\right)$

11. $-5\frac{1}{4} \div \frac{7}{8}$

12. $8\frac{1}{3} \div \left(-\frac{5}{9}\right)$

13. Evaluate $\frac{x}{y}$ if $x = \frac{4}{5}$ and $y = -0.1$. Write your answer in simplest form. **(Example 3)**

15. Evaluate $\frac{c}{d}$ if $c = -4.75$ and $d = -1\frac{1}{4}$. Write your answer in simplest form. **(Example 3)**

Test Practice

14. Equation Editor Evaluate $\frac{a}{b}$ if $a = -\frac{1}{4}$ and $b = 0.02$. Write your answer in simplest form.

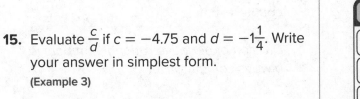

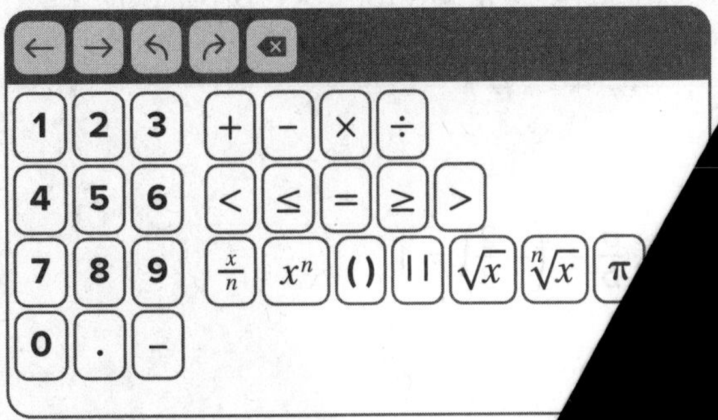

Apply

16. The photography club is selling hot chocolate at soccer games to raise money for new cameras. The table shows their profit per game for the first five games. Based on the average profit per game, how much total money can the club expect to earn by the end of the 10-game season?

Game	Profit ($)
1	−12.50
2	−10.15
3	18.65
4	25.90
5	45.75

17. The change in the total amount of Suki's allowance balance over the course of four weeks is shown in the table. Based on the average change in her allowance balance per week, how much money can Suki expect to save or spend over the course of a year?

Week	Change ($)
1	25.25
2	−5.75
3	12.40
4	−14.35

18. MP **Find the Error** A student is finding $-\frac{6}{7} \div \left(-\frac{5}{6}\right)$. Find the student's mistake and correct it.

$$-\frac{6}{7} \div \left(-\frac{5}{6}\right) = -\frac{7}{6} \cdot \left(-\frac{5}{6}\right)$$
$$= \frac{35}{36}$$

19. MP **Make an Argument** Which is greater, $20 \cdot \frac{3}{4}$ or $20 \div \frac{3}{4}$? Explain.

20. MP **Persevere with Problems** Find the missing fraction for each problem.

$\ldots \div \left(\frac{x}{y}\right) = -\frac{1}{2}$

21. MP **Use a Counterexample** Is the following statement *true* or *false*? If false, provide a counterexample.

Division of negative rational numbers is always commutative.

Apply Rational Number Operations

I Can... add, subtract, multiply, and divide rational numbers, including using those four operations to solve real-world problems.

Learn Apply Rational Number Operations

The order of operations used to simplify numerical expressions with whole numbers also applies to simplifying numerical expressions with rational numbers.

If the expression contains fractions and decimals, use the Commutative Properties or the Associative Properties to group like forms together. Simplify as much as possible, and then rename the numbers in the same form.

Example 1 Apply Rational Number Operations

Evaluate $ab + c - d$ if $a = \frac{5}{6}$, $b = -\frac{4}{5}$, $c = 0.75$, and $d = \frac{1}{3}$.

Substitute the values into the equation.

$$\frac{5}{6} \cdot \left(-\frac{4}{5}\right) + 0.75 - \frac{1}{3}$$

$\frac{5}{6} \cdot \left(-\frac{4}{5}\right) + 0.75 - \frac{1}{3} = -\frac{2}{3} + 0.75 - \frac{1}{3}$	Multiply the fractions.
$= -\frac{2}{3} + 0.75 + \left(-\frac{1}{3}\right)$	Add the additive inverse.
$= -\frac{2}{3} + \left(-\frac{1}{3}\right) + 0.75$	Commutative Property
$= -1 + 0.75$	Add the fractions.
$= -0.25$	Add.

So, the value of the expression is ________.

Think About It!

How will you use the order of operations to evaluate this expression?

Talk About It!

Was it helpful to use the Commutative Property to change the order? Explain.

Your Notes

Check

Evaluate $a + bc + d$ if $a = -3.6$, $b = 2\frac{4}{5}$, $c = -\frac{1}{2}$, and $d = 6\frac{3}{4}$.

Show your work here

Example 2 Apply Rational Number Operations

Evaluate $\frac{2}{5}(x + y) + z$ if $x = \frac{7}{8}$, $y = -\frac{1}{4}$, and $z = -\frac{2}{3}$.

Substitute the values into the equation.

$$\square\left[\square + (\square)\right] + (\square)$$

Simplify the expression.

$\frac{2}{5} \times \left[\frac{7}{8} + \left(-\frac{1}{4}\right)\right] + \left(-\frac{2}{3}\right)$

$= \frac{2}{5} \times \left[\frac{7}{8} + \left(-\frac{2}{8}\right)\right] + \left(-\frac{2}{3}\right)$ Write $-\frac{1}{4}$ as $-\frac{2}{8}$.

$= \frac{2}{5}\left(\square\right) + \left(-\frac{2}{3}\right)$ Add.

$= \left(\square\right) + \left(-\frac{2}{3}\right)$ Multiply.

$= \frac{3}{12} + \left(-\frac{8}{12}\right)$ Find a common denominator.

$= \left(\square\right)$ Add.

So, the value of the expression is $-\frac{5}{12}$.

Think About It! How would you begin evaluating this expression?

Talk About It! Why would it be easier to do the operation in the parentheses rather than using the Distributive Property to solve the example?

Check

Evaluate $x + y(0.4 + z)$ if $x = -2.6$, $y = -\frac{1}{2}$, and $z = -\frac{5}{8}$.

Show your work here

Go Online You can complete an Extra Example online.

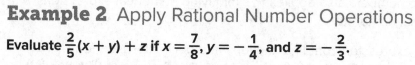

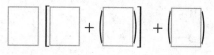

Apply Food

Aarya has a recipe for blonde brownies. She doesn't like a lot of walnuts, but loves chocolate, so she is using $\frac{1}{2}$ of the amount of walnuts called for and increasing the chocolate chips by $\frac{1}{3}$. She then wants to double her new recipe. How many cups of walnuts will Aarya need? How many cups of chocolate chips?

Quantity	Ingredient
$\frac{1}{2}$ cup	chopped walnuts
$\frac{3}{4}$ cup	chocolate chips

Go Online Watch the animation.

1 What is the task?

Make sure you understand exactly what question to answer or problem to solve. You may want to read the problem three times. Discuss these questions with a partner.

First Time Describe the context of the problem, in your own words.
Second Time What mathematics do you see in the problem?
Third Time What are you wondering about?

2 How can you approach the task? What strategies can you use?

Record your observations here

Talk About It! How is finding the amount of walnuts for the recipe different than finding the amount of chocolate chips?

3 What is your solution?

Use your strategy to solve the problem.

Show your work here

4 How can you show your solution is reasonable?

Write About It! Write an argument that can be used to defend your solution.

Check

Adel City is enclosing a skate park with fencing. The skate park is $31\frac{1}{3}$ yards wide and $24\frac{2}{3}$ yards long. Fencing is sold in 8-foot sections and costs $67.99 per section. How much will it cost to fence in the entire skate park?

Show your work here

Go Online You can complete an Extra Example online.

Pause and Reflect

Why is it helpful to use parentheses when substituting values for variables?

Record your observations here

Name ________________ Period ________ Date ________

Practice

Go Online You can complete your homework online.

Evaluate each expression if $a = \frac{7}{8}$, $b = -\frac{7}{16}$, $c = 0.8$, $d = \frac{1}{4}$. (Example 1)

1. $\frac{a}{b} - c + d$

2. $a + b + cd$

3. $c - a + \frac{b}{d}$

4. $\frac{b}{a} + \frac{c}{d}$

5. $d^2 - b$

6. $a + \frac{b}{c} - d$

Evaluate each expression if $a = \frac{7}{10}$, $b = \frac{3}{5}$, $c = -1.9$, and $d = -\frac{1}{5}$. Write your answer in simplest form. (Example 2)

7. $\frac{1}{2}(a - b) + c$

8. $0.25(a + b) - d$

9. $b + 0.2(c - d)$

10. $(b + d) - a$

11. $b^2 + \frac{1}{10}(d - b)$

12. $b + \frac{c}{d}$

Test Practice

13. Equation Editor Evaluate $4.5(xy) - \frac{x}{5}$ if $x = 12$ and $y = \frac{2}{3}$.

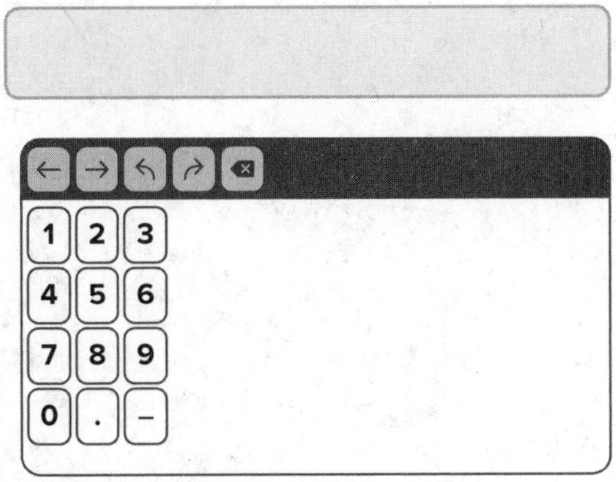

Apply

14. Jake is enclosing his vegetable garden with fencing. The table shows the dimensions of his rectangular garden. Fencing is sold in 2.5-foot sections and costs \$25.99 per section. How much will it cost to fence in the entire garden?

Garden Dimension	Measurement (yards)
Length	$14\frac{2}{3}$
Width	$10\frac{1}{3}$

15. Maya has a recipe for blueberry chocolate chip muffins. She doesn't like a lot of oats, but loves blueberries, so she is using $\frac{3}{4}$ of the amount of oats called for and increasing the amount of blueberries by $\frac{1}{4}$ of the original amount. She then wants to double her new recipe. How many cups of oats will Maya need? How many cups of blueberries?

Ingredient	Amount (cups)
Blueberries	$\frac{3}{4}$
Chocolate Chips	1
Oats	$\frac{1}{2}$
White Sugar	1

16. MP **Justify Conclusions** A student says that the Commutative Property and Associative Property cannot be used to simplify expressions with rational numbers. Is the student correct? Justify your answer.

17. MP **Persevere with Problems** There are 120 bouncy balls in a display bin. Of the bouncy balls in the bin, one-half are priced at \$0.25 each, one-fifth are priced at \$0.75, and the remaining bouncy balls are priced at \$1.50. What is the total value of the bouncy balls in the bin? Write a numerical expression that can be used to solve the problem. Then solve it.

18. A preschool teacher has 17.5 feet of yarn for her students to make necklaces to raise money for a local animal shelter. Each necklace requires 14 inches of yarn. If the necklaces sell for \$2.50, how much money will they raise if they sell all the necklaces? Explain how you solved.

19. Create Write and solve a multi-step real-world problem where you apply rational number operations.

📖 **Foldables** Use your Foldable to help review the module.

Operations With Rational Numbers

Tab 1

Rule

Rule

Rule

Rule

Tab 2

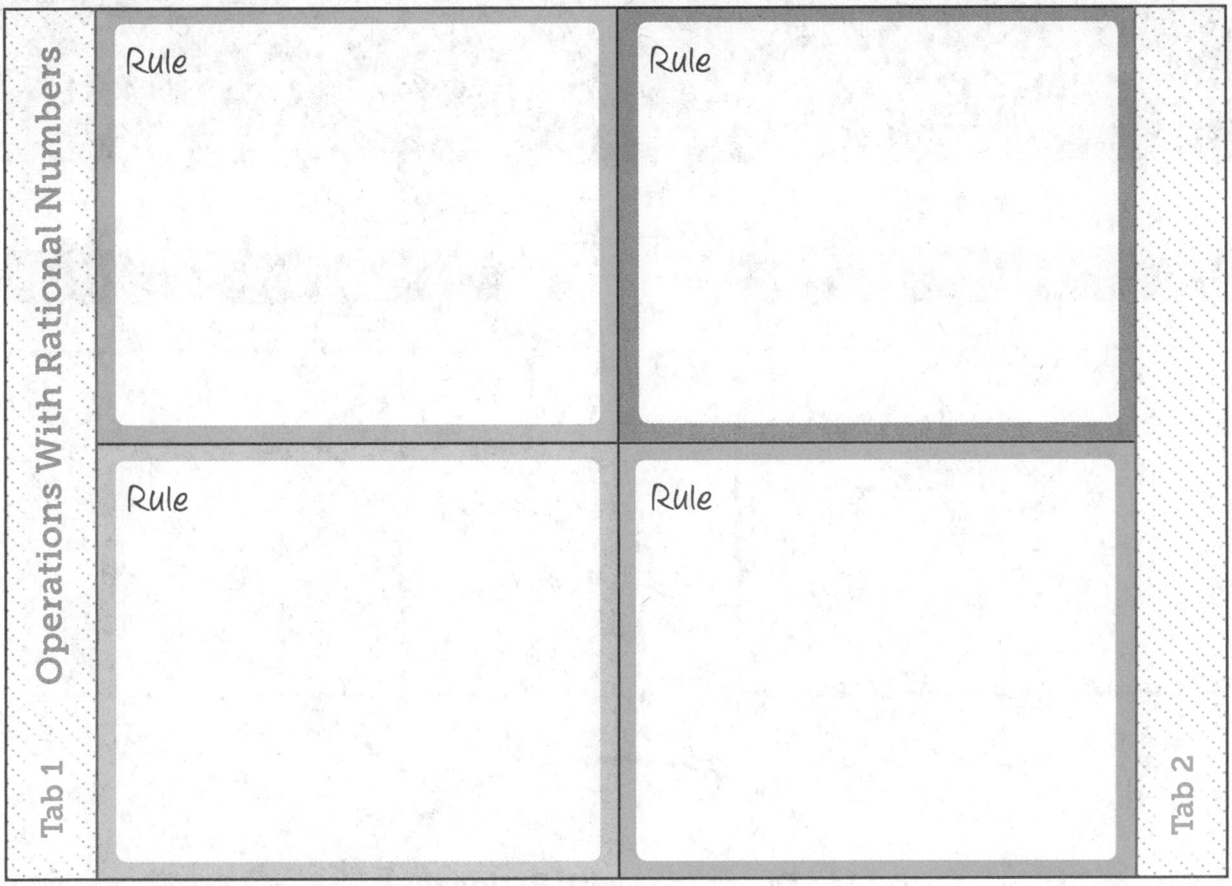

Rate Yourself! ⬛ ◆ ★

Complete the chart at the beginning of the module by placing a checkmark in each row that corresponds with how much you know about each topic after completing this module.

Write about one thing you learned.

Write about a question you still have.

Reflect on the Module

Use what you learned about operations with rational numbers to complete the graphic organizer.

Essential Question

How are operations with rational numbers related to operations with integers?

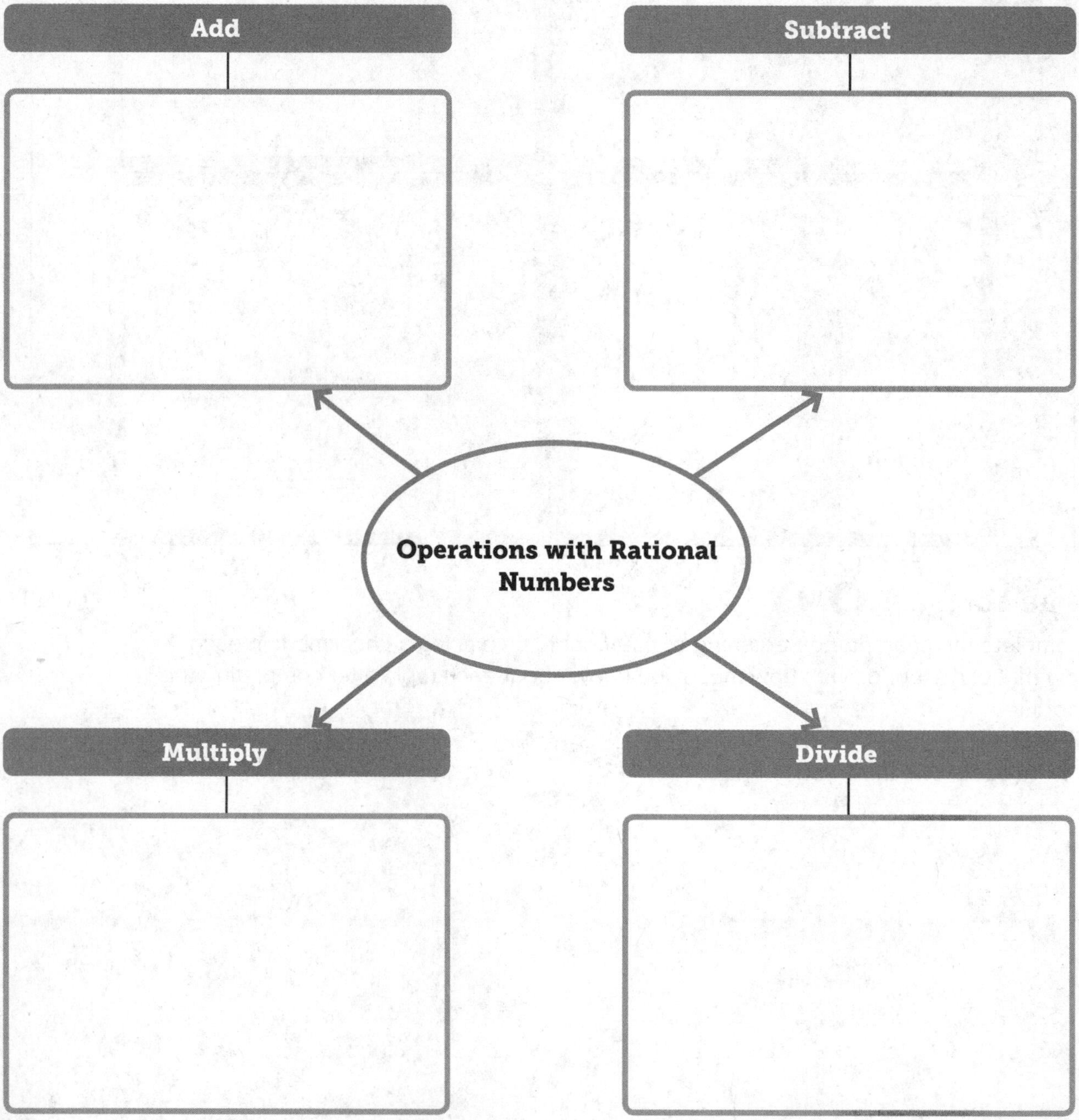

Name _______________ Period _______ Date _______

Test Practice

1. Equation Editor There were 20 questions on the last quiz in Mrs. Lorenzo's class. A student answered 17 questions correctly. Write the ratio of correct answers to total questions as a decimal. **(Lesson 1)**

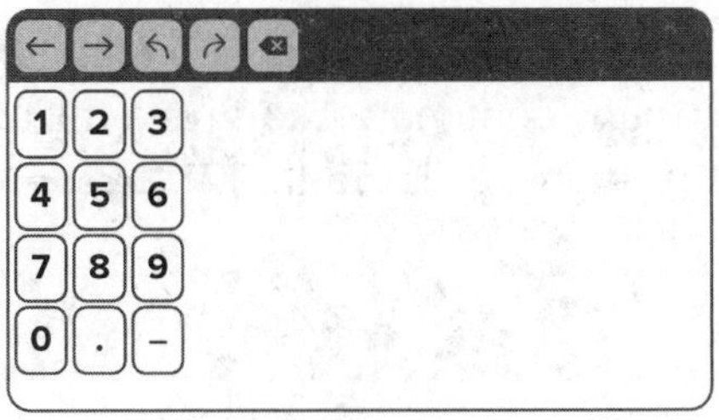

2. Multiple Choice Which of the following is the decimal representation of the rational number $-\frac{5}{12}$? **(Lesson 1)**

Ⓐ $-0.41\overline{6}$

Ⓑ $0.41\overline{6}$

Ⓒ $-0.\overline{416}$

Ⓓ $0.\overline{416}$

3. Table Item Jeffrey earned $23.50 mowing lawns and $17.25 babysitting. **(Lesson 2)**

A. Find the additive inverse of the total amount that Jeffrey earned.

B. Place a ✓ in the "yes" column for the situations that could represent the additive inverse of how much Jeffrey earned.

	yes	no
losing $40.75		
donating $40.75		
finding $40.75		
earning $40.75		
spending $40.75		
receiving a gift of $40.75		

4. Open Response The manager of Happy Puppy dog boarding company begins the month with 85.4 pounds of dog food and tracks the weekly use, along with the supply delivery received in the middle of the month. **(Lesson 2)**

Week	Weekly Dog Food Use (lb)
1	−28.4
2	$-21\frac{5}{8}$
3	$97\frac{3}{4}$
4	−30.25

How much dog food does the company have on hand at the end of the month?

5. Equation Editor At the beginning of the day, Meredith has 85.4 pounds of apples to sell at a farmer's market. The table shows the hourly change in the amount of apples she has after the first several hours. **(Lesson 3)**

Hour	Change in Apples (lb)
1	−8.5
2	$-12\frac{2}{3}$
3	$-16\frac{1}{5}$
4	−10.2

After 4 hours, about how many pounds of apples does Meredith have left?

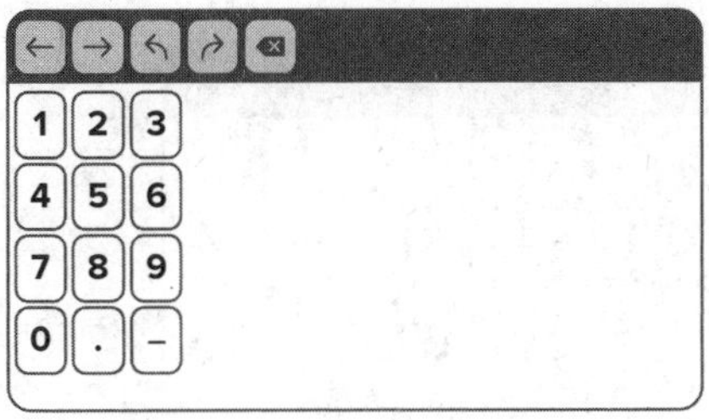

6. Multiple Choice What is the result when $-\frac{7}{8}$ is multiplied by -3.5? **(Lesson 4)**

Ⓐ $3\frac{1}{16}$

Ⓑ $2\frac{11}{16}$

Ⓒ $-2\frac{11}{16}$

Ⓓ $-3\frac{1}{16}$

7. Open Response The area of a triangle is found by evaluating the expression $\frac{1}{2}bh$, where b is the base of the triangle and h is the height. Find the area of the triangle shown. Explain how you know your answer is reasonable. **(Lesson 4)**

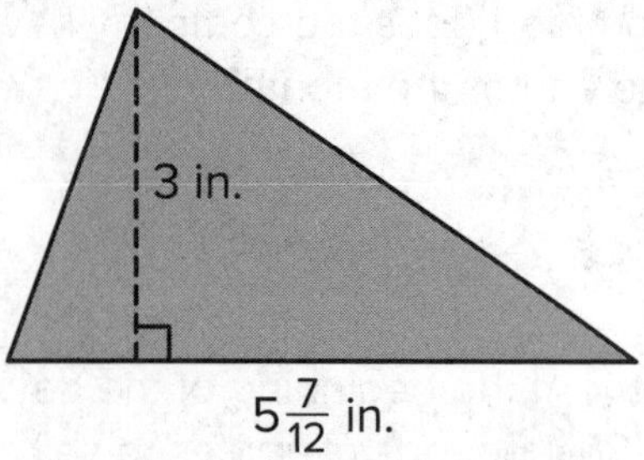

8. Equation Editor Find $-\frac{11}{12} \div \left(-\frac{2}{3}\right)$. Express your answer as a fraction or mixed number in simplest form. **(Lesson 5)**

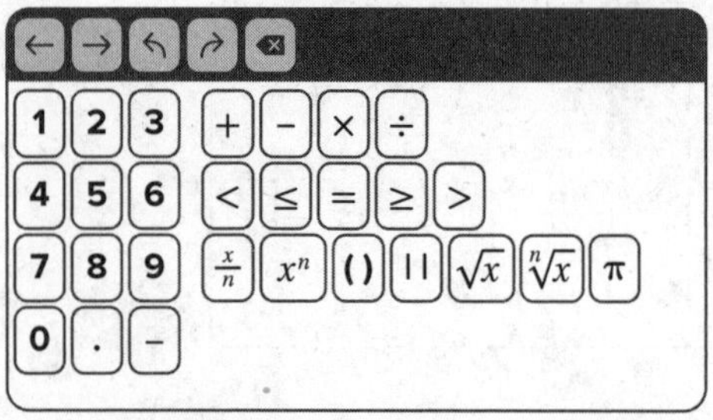

9. Multiple Choice What is the simplified value of the expression $\frac{m}{n}$ if $m = \frac{4}{5}$ and $n = -0.25$? **(Lesson 5)**

Ⓐ $3\frac{1}{5}$

Ⓑ $\frac{8}{25}$

Ⓒ $-\frac{8}{25}$

Ⓓ $-3\frac{1}{5}$

10. Equation Editor A U.S. nickel is 1.95 mm thick. How many centimeters long is a roll of nickels if the entire roll is worth $2? **(Lesson 6)**

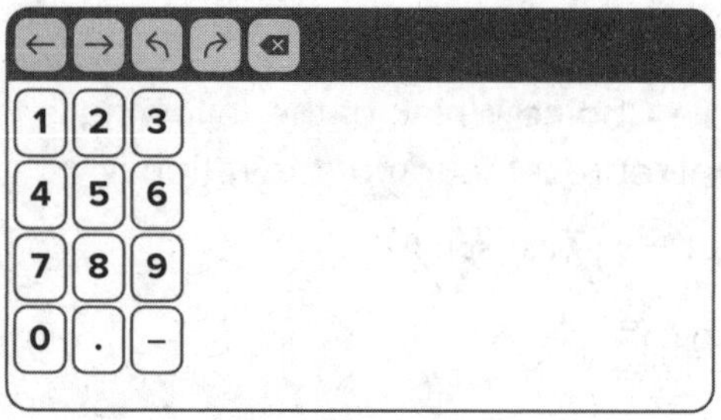

11. Open Response Nathaniel has 132 DVDs in his collection. Each DVD case has the dimensions shown. He plans to buy shelves for the collection that are 31.5 inches wide and cost $15.75 each. How much will it cost Nathaniel to shelve his DVD collection? Explain how you found your answer. **(Lesson 6)**

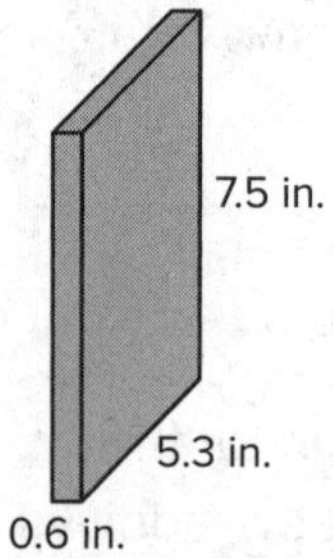

Foldables Study Organizers

What Are Foldables and How Do I Create Them?

Foldables are three-dimensional graphic organizers that help you create study guides for each module in your book.

Step 1 Go to the back of your book to find the Foldable for the module you are currently studying. Follow the cutting and assembly instructions at the top of the page.

Step 2 Go to the Module Review at the end of the module you are currently studying. Match up the tabs and attach your Foldable to this page. Dotted tabs show where to place your Foldable. Striped tabs indicate where to tape the Foldable.

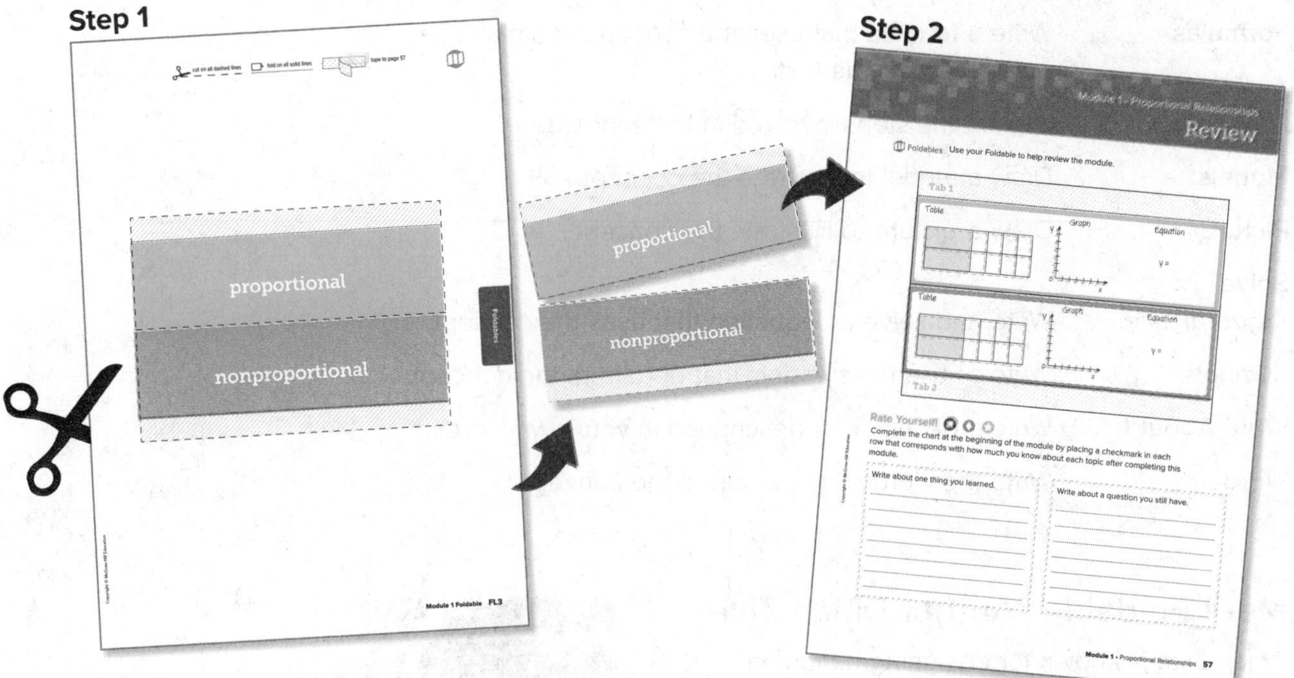

How Will I Know When to Use My Foldable?

You will be directed to work on your Foldable at the end of selected lessons. This lets you know that it is time to update it with concepts from that lesson. Once you've completed your Foldable, use it to study for the module test.

How Do I Complete My Foldable?

No two Foldables in your book will look alike. However, some will ask you to fill in similar information. Below are some of the instructions you'll see as you complete your Foldable. **HAVE FUN** learning math using Foldables!

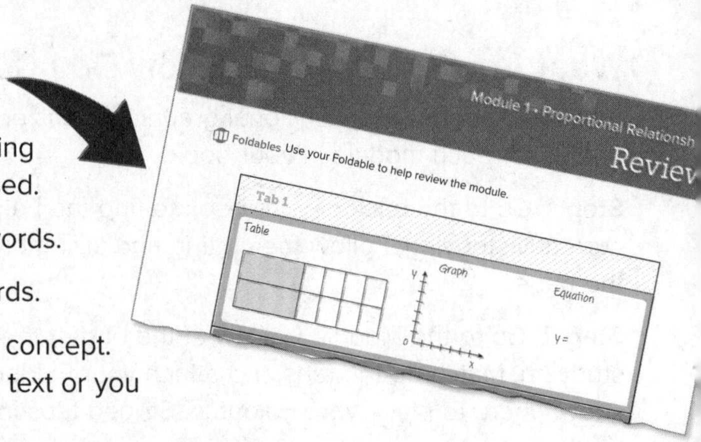

Instructions and What They Mean

Best Used to...	Complete the sentence explaining when the concept should be used.
Definition	Write a definition in your own words.
Description	Describe the concept using words.
Equation	Write an equation that uses the concept. You may use one already in the text or you can make up your own.
Example	Write an example about the concept. You may use one already in the text or you can make up your own.
Formulas	Write a formula that uses the concept. You may use one already in the text.
How do I ...?	Explain the steps involved in the concept.
Models	Draw a model to illustrate the concept.
Picture	Draw a picture to illustrate the concept.
Solve Algebraically	Write and solve an equation that uses the concept.
Symbols	Write or use the symbols that pertain to the concept.
Write About It	Write a definition or description in your own words.
Words	Write the words that pertain to the concept.

Meet Foldables Author Dinah Zike

Dinah Zike is known for designing hands-on manipulatives that are used nationally and internationally by teachers and parents. Dinah is an explosion of energy and ideas. Her excitement and joy for learning inspires everyone she touches.

cut on all dashed lines   fold on all solid lines   tape to page 57

proportional

nonproportional

Foldables

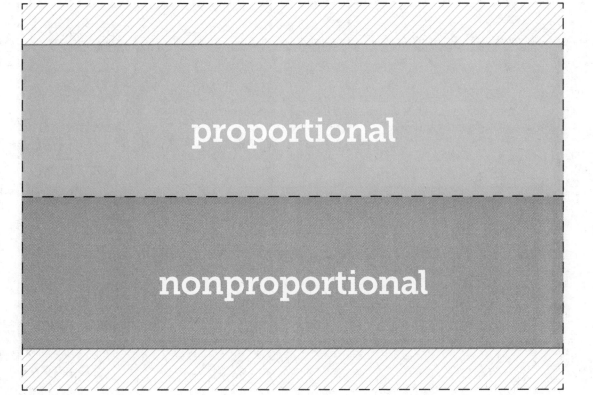

cut on all dashed lines   fold on all solid lines   tape to page 57

Tab 1

Write About it

Write About it

Tab 2

cut on all dashed lines   fold on all solid lines   tape to page 121

Percents

percent of increase

percent of decrease

Foldables

cut on all dashed lines

fold on all solid lines

tape to page 121

Definition

Definition

cut on all dashed lines

fold on all solid lines

tape to page 171

Operations with Integers

add

subtract

multiply

divide

Foldables

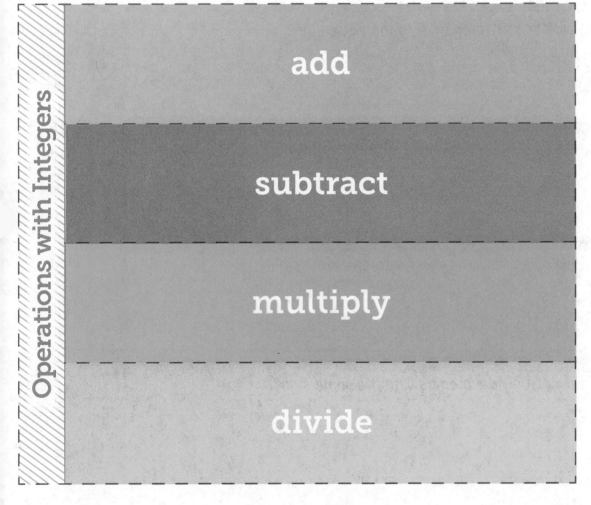

cut on all dashed lines

fold on all solid lines

tape to page 171

How do I add integers with the same sign?

+

How do I subtract integers with the same sign?

−

How do I multiply integers with the same sign?

×

How do I divide integers with the same sign?

÷

cut on all dashed lines

fold on all solid lines

tape to page 227

Operations with Rational Numbers

+ or − like rational numbers	÷ rational numbers
× rational numbers	+ or − unlike rational numbers

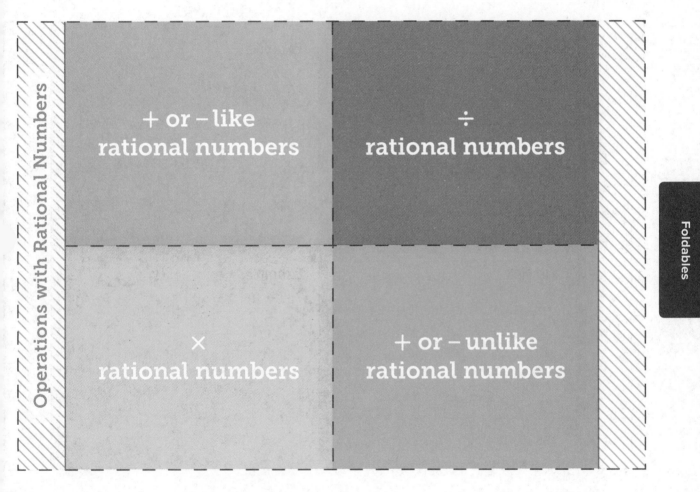

cut on all dashed lines fold on all solid lines tape to page 227

Examples	Examples
Examples	Examples

Tab 2

Tab 1

Glossary

The eGlossary contains words and definitions in the following 14 languages:

Arabic	**English**	**Hmong**	**Russian**	**Urdu**
Bengali	**French**	**Korean**	**Spanish**	**Vietnamese**
Brazilian Portuguese	**Haitian Creole**	**Mandarin**	**Tagalog**	

English | Español

A

absolute value (Lesson 3-1) The distance the number is from zero on a number line.

valor absoluto Distancia a la que se encuentra un número de cero en la recta numérica.

acute angle (Lesson 8-1) An angle with a measure greater than 0° and less than 90°.

ángulo agudo Ángulo que mide más de 0° y menos de 90°.

acute triangle (Lesson 8-3) A triangle having three acute angles.

triángulo acutángulo Triángulo con tres ángulos agudos.

Addition Property of Equality (Lesson 6-1) If you add the same number to each side of an equation, the two sides remain equal.

propiedad de adición de la igualdad Si sumas el mismo número a ambos lados de una ecuación, los dos lados permanecen iguales.

Addition Property of Inequality (Lesson 7-1) If you add the same number to each side of an inequality, the inequality remains true.

propiedad de desigualdad en la suma Si se suma el mismo número a cada lado de una desigualdad, la desigualdad sigue siendo verdadera.

Additive Identity Property (Lesson 3-1) The sum of any number and zero is the number.

propiedad de identidad de la suma La suma de cualquier número y cero es el mismo número.

additive inverse (Lesson 3-1) Two integers that are opposites. The sum of an integer and its additive inverse is zero.

inverso aditivo Dos enteros opuestos. La suma de un entero y su inverso aditiva es cero.

Additive Inverse Property (Lesson 3-1) The sum of any number and its additive inverse is zero.

propiedad inversa aditiva La suma de cualquier número y su inversa aditiva es cero.

adjacent angles (Lesson 8-1) Angles that have the same vertex, share a common side, and do not overlap.

ángulos adyacentes Ángulos que comparten el mismo vértice y un común lado, pero no se sobreponen.

algebra (Lesson 3-1) A branch of mathematics that involves expressions with variables.

álgebra Rama de las matemáticas que trata de las expresiones con variables.

algebraic expression (Lesson 5-1) A combination of variables, numbers, and at least one operation.

expresión algebraica Combinación de variables, números y por lo menos una operación.

amount of error (Lesson 2-4) The positive difference between the estimate and the actual amount.

cantidad de error La diferencia positiva entre la estimación y la cantidad real.

angle (Lesson 8-1) Two rays with a common endpoint form an angle. The rays and vertex are used to name the angle.

ángulo Dos rayos con un extremo común forman un ángulo. Los rayos y el vértice se usan para nombrar el ángulo.

area (Lesson 9-2) The measure of the interior surface of a two-dimensional figure.

área La medida de la superficie interior de una figura bidimensional.

asymmetric distribution (Lesson 11-4) A distribution in which the shape of the graph on one side of the center is very different than the other side, or it has outliers that might affect the average.

distribución asimétrica Una distribución en la que la forma del gráfico en un lado del centro es muy diferente del otro lado, o tiene valores atípicos que pueden afectar al promedio.

B

bar notation (Lesson 4-1) In repeating decimals, the line or bar placed over the digits that repeat. For example, $2.\overline{63}$ indicates that the digits 63 repeat.

notación de barra Línea o barra que se coloca sobre los dígitos que se repiten en decimales periódicos. Por ejemplo, $2.\overline{63}$ indica que los dígitos 63 se repiten.

base (Lesson 8-5) One of the two parallel congruent faces of a prism.

base Una de las dos caras paralelas congruentes de un prisma.

biased sample (Lesson 11-1) A sample drawn in such a way that one or more parts of the population are favored over others.

muestra sesgada Muestra en que se favorece una o más partes de una población.

box plot (Lesson 11-4) A method of visually displaying a distribution of data values by using the median, quartiles, and extremes of the data set. A box shows the middle 50% of the data.

diagrama de caja Un método de mostrar visualmente una distribución de valores usando la mediana, cuartiles y extremos del conjunto de datos. Una caja muestra el 50% del medio de los datos.

C

center (Lesson 9-1) The point from which all points on a circle are the same distance.

centro El punto desde el cual todos los puntos en una circunferencia están a la misma distancia.

circle (Lesson 9-1) The set of all points in a plane that are the same distance from a given point called the center.

círculo Conjunto de todos los puntos de un plano que están a la misma distancia de un punto dado denominado centro.

circumference (Lesson 9-1) The distance around a circle.

circunferencia Distancia en torno a un círculo.

coefficient (Lesson 5-1) The numerical factor of a term that contains a variable.

coeficiente El factor numérico de un término que contiene una variable.

commission (Lesson 2-5) A payment equal to a percent of the amount of goods or services that an employee sells for the company.

comisión Un pago igual a un porcentaje de la cantidad de bienes o servicios que un empleado vende para la empresa.

common denominator (Lesson 4-1) A common multiple of the denominators of two or more fractions. 24 is a common denominator for $\frac{1}{3}$, $\frac{5}{8}$, and $\frac{3}{4}$ because 24 is the LCM of 3, 8, and 4.

común denominador El múltiplo común de los denominadores de dos o más fracciones. 24 es un denominador común para $\frac{1}{3}$, $\frac{5}{8}$, y $\frac{3}{4}$ porque 24 es el mcm de 3, 8 y 4.

complementary angles (Lesson 8-2) Two angles are complementary if the sum of their measures is 90°.

ángulos complementarios Dos ángulos son complementarios si la suma de sus medidas es 90°.

complementary events (Lesson 10-3) Two events in which either one or the other must happen, but they cannot happen at the same time. The sum of the probability of an event and its complement is 1 or 100%.

eventos complementarios Dos eventos en los cuales uno o el otro debe suceder, pero no pueden ocurrir al mismo tiempo. La suma de la probabilidad de un evento y su complemento es 1 o 100%.

complex fraction (Lesson 1-1) A fraction $\frac{A}{B}$ where A and/or B are fractions and B does not equal zero.

fracción compleja Una fracción $\frac{A}{B}$ en la cual A y/o B son fracciones y B no es igual a cero.

composite figure (Lesson 9-3) A figure that is made up of two or more figures.

figura compuesta Figura formada por dos o más figuras.

compound event (Lesson 10-5) An event consisting of two or more simple events.

evento compuesto Un evento que consiste en dos o más eventos simples.

cone (Lesson 8-5) A three-dimensional figure with one circular base connected by a curved surface to a single point.

cono Una figura tridimensional con una base circular conectada por una superficie curva para un solo punto.

congruent (Lesson 8-1) Having the same measure.

congruente Que tiene la misma medida.

congruent angles (Lesson 8-1) Angles that have the same measure.

ángulos congruentes Ángulos que tienen la misma medida.

congruent figures (Lesson 8-3) Figures that have the same size and same shape and corresponding sides and angles with equal measure.

figuras congruentes Figuras que tienen el mismo tamaño y la misma forma y los lados y los ángulos correspondientes tienen igual medida.

congruent segments (Lesson 8-3) Sides with the same length.

segmentos congruentes Lados con la misma longitud.

constant (Lesson 5-1) A term that does not contain a variable.

constante Término que no contiene ninguna variable.

constant of proportionality (Lesson 1-3) A constant ratio or unit rate of two variable quantities. It is also called the constant of variation.

constante de proporcionalidad Una razón constante o tasa por unidad de dos cantidades variables. También se llama constante de variación.

constant of variation (Lesson 1-3) The constant ratio in a direct variation. It is also called the constant of proportionality.

constante de variación Una razón constante o tasa por unidad de dos cantidades variables. También se llama constante de proporcionalidad.

constant rate of change (Lesson 1-3) The rate of change in a linear relationship.

razón constante de cambio Tasa de cambio en una relación lineal.

convenience sample (Lesson 11-1) A sample which consists of members of a population that are easily accessed.

muestra de conveniencia Muestra que incluye miembros de una población fácilmente accesibles.

cross section (Lesson 8-5) The intersection of a solid and a plane.

sección transversal Intersección de un sólido con un plano.

cylinder (Lesson 8-5) A three-dimensional figure with two parallel congruent circular bases connected by a curved surface.

cilindro Una figura tridimensional con dos paralelas congruentes circulares bases conectados por una superficie curva.

D

defining a variable (Lesson 6-1) Choosing a variable and a quantity for the variable to represent in an expression or equation.

definir una variable El eligir una variable y una cantidad que esté representada por la variable en una expresión o en una ecuacion.

degrees (Lesson 8-1) The most common unit of measure for angles. If a circle were divided into 360 equal-sized parts, each part would have an angle measure of 1 degree.

grados La unidad más común para medir ángulos. Si un círculo se divide en 360 partes iguales, cada parte tiene una medida angular de 1 grado.

diameter (Lesson 9-1) The distance across a circle through its center.

diámetro Segmento que pasa por el centro de un círculo y lo divide en dos partes iguales.

dimensional analysis (Lesson 1-2) The process of including units of measurement when you compute.

análisis dimensional Proceso que incluye las unidades de medida al hacer cálculos.

discount (Lesson 2-8) The amount by which the regular price of an item is reduced.

descuento Cantidad que se le rebaja al precio regular de un artículo.

distribution (Lesson 11-4) The shape of a graph of data.

distribución La forma de un gráfico de datos.

Distributive Property (Lesson 3-3) To multiply a sum by a number, multiply each addend of the sum by the number outside the parentheses. For any numbers a, b, and c, $a(b + c) = ab + ac$ and $a(b - c) = ab - ac$.

Example: $2(5 + 3) = (2 \times 5) + (2 \times 3)$ and $2(5 - 3) = (2 \times 5) - (2 \times 3)$

propiedad distributiva Para multiplicar una suma por un número, multiplíquese cada sumando de la suma por el número que está fuera del paréntesis. Sean cuales fuere los números a, b, y c, $a(b + c) = ab + ac$ y $a(b - c) = ab - ac$.

Ejemplo: $2(5 + 3) = (2 \cdot 5) + (2 \cdot 3)$ y $2(5 - 3) = (2 \cdot 5) - (2 \cdot 3)$

Division Property of Equality (Lesson 6-1) If you divide each side of an equation by the same nonzero number, the two sides remain equal.

propiedad de igualdad de la división Si divides ambos lados de una ecuación entre el mismo número no nulo, los lados permanecen iguales.

Division Property of Inequality (Lesson 7-3) When you divide each side of an inequality by a negative number, the inequality symbol must be reversed for the inequality to remain true.

propiedad de desigualdad en la división Cuando se divide cada lado de una desigualdad entre un número negativo, el símbolo de desigualdad debe invertirse para que la desigualdad siga siendo verdadera.

double box plot **(Lesson 11-4)** Two box plots graphed on the same number line.

doble diagrama de caja Dos diagramas de caja sobre la misma recta numérica.

double line plot **(Lesson 11-4)** A method of visually displaying a distribution of two sets of data values where each value is shown as a dot above a number line.

doble diagrama de línea Un método de mostrar visualmente una distribución de dos conjuntos de valores donde cada valor se muestra como un punto arriba de una recta numérica.

E

edge **(Lesson 8-5)** The line segment where two faces of a polyhedron intersect.

borde El segmento de línea donde se cruzan dos caras de un poliedro.

enlargement **(Lesson 8-4)** An image larger than the original.

ampliación Imagen más grande que la original.

equation **(Lesson 6-1)** A mathematical sentence that contains an equals sign, =, stating that two quantities are equal.

ecuación Enunciado matemático que contiene el signo de igualdad = indicando que dos cantidades son iguales.

equiangular **(Lesson 8-3)** In a polygon, all of the angles are congruent.

equiangular En un polígono, todos los ángulos son congruentes.

equilateral **(Lesson 8-3)** In a polygon, all of the sides are congruent.

equilátero En un polígono, todos los lados son congruentes.

equilateral triangle **(Lesson 8-3)** A triangle having three congruent sides.

triángulo equilátero Triángulo con tres lados congruentes.

equivalent equations **(Lesson 6-1)** Two or more equations with the same solution.

ecuaciones equivalentes Dos o más ecuaciones con la misma solución.

equivalent expressions **(Lesson 5-1)** Expressions that have the same value.

expresiones equivalentes Expresiones que tienen el mismo valor.

equivalent ratios **(Lesson 1-2)** Two ratios that have the same value.

razones equivalentes Dos razones que tienen el mismo valor.

evaluate **(Lesson 6-1)** To find the value of an expression.

evaluar Calcular el valor de una expresión.

event **(Lesson 10-1)** The desired outcome or set of outcomes in a probability experiment.

evento El resultado deseado o conjunto de resultados en un experimento de probabilidad.

experimental probability **(Lesson 10-2)** An estimated probability based on the relative frequency of positive outcomes occurring during an experiment. It is based on what *actually* occurred during such an experiment.

probabilidad experimental Probabilidad estimada que se basa en la frecuencia relativa de los resultados positivos que ocurren durante un experimento. Se basa en lo que en *realidad* ocurre durante dicho experimento.

F

face (Lesson 8-5) A flat surface of a polyhedron.

cara Una superficie plana de un poliedro.

factor (Lesson 5-4) To write a number as a product of its factors.

factorizar Escribir un número como el producto de sus factores.

factored form (Lesson 5-4) An expression expressed as the product of its factors.

forma factorizada Una expresión expresada como el producto de sus factores.

factors (Lesson 5-1) Two or more numbers that are multiplied together to form a product.

factores Dos o más números que se multiplican entre sí para formar un producto.

fee (Lesson 2-5) A payment for a service. It can be a fixed amount, a percent of the charge, or both.

cuota Un pago por un servicio. Puede ser una cantidad fija, un porcentaje del cargo, o ambos.

G

graph (Lesson 1-4) The process of placing a point on a number line or on a coordinate plane at its proper location.

graficar Proceso de dibujar o trazar un punto en una recta numérica o en un plano de coordenadas en su ubicación correcta.

gratuity (Lesson 2-7) Also known as a tip. It is a small amount of money in return for a service.

gratificación También conocida como propina. Es una cantidad pequeña de dinero en retribución por un servicio.

greatest common factor (GCF) of two monomials (Lesson 5-4) The greatest monomial that is a factor of both monomials. The greatest common factor also includes any variables that the monomials have in common.

mayor factor común (GCF) de dos monomios El monomio más grande que es un factor de ambos monomios. El factor común más grande también incluye las variables que los monomios tienen en común.

I

inequality (Lesson 7-1) An open sentence that uses $<, >, \neq, \leq$, or $\geq$ to compare two quantities.

desigualdad Enunciado abierto que usa $<, >, \neq, \leq$, o $\geq$ para comparar dos cantidades.

inference (Lesson 11-1) A prediction made about a population.

inferencia Una predicción hecha sobre una población.

integer (Lesson 3-1) Any number from the set $\{..., -4, -3, -2, -1, 0, 1, 2, 3, 4, ...\}$, where ... means continues without end.

entero Cualquier número del conjunto $\{..., -4, -3, -2, -1, 0, 1, 2, 3, 4, ...\}$, donde ... significa que continúa sin fin.

interest (Lesson 2-9) The amount paid or earned for the use of the principal.

interés La cantidad pagada o ganada por el uso del principal.

interquartile range (IQR) (Lesson 11-4) A measure of variation in a set of numerical data, the interquartile range is the distance between the first and third quartiles of the data set.

rango intercuartil (RIQ) El rango intercuartil, una medida de la variacion en un conjunto de datos numéricos, es la distancia entre el primer y el tercer cuartil del conjunto de datos

invalid inference (Lesson 11-1) An inference that is based on a biased sample or makes a conclusion not supported by the results of the sample.

inferencia inválida Una inferencia que se basa en una muestra sesgada o hace una conclusión no apoyada por los resultados de la muestra.

Inverse Property of Multiplication (Lesson 6-1) The product of a number and its multiplicative inverse is 1.

propiedad inversa de la multiplicación El producto de un número y su inverso multiplicativo es 1.

isosceles triangle (Lesson 8-3) A triangle having at least two congruent sides.

triángulo isósceles Triángulo que tiene por lo menos dos lados congruentes.

L

lateral face (Lesson 9-5) In a polyhedron, a face that is not a base.

cara lateral En un poliedro, las caras que no forman las bases.

lateral surface area (Lesson 9-5) The sum of the areas of all of the lateral faces of a solid.

área de superficie lateral Suma de las áreas de todas las caras de un sólido.

least common denominator (LCD) (Lesson 4-1) The least common multiple of the denominators of two or more fractions. You can use the LCD to compare fractions.

mínimo común denominador (mcd) El menor de los múltiplos de los denominadores de dos o más fracciones. Puedes usar el mínimo común denominador para comparar fracciones.

like terms (Lesson 5-1) Terms that contain the same variable(s) raised to the same power. Example: $5x$ and $6x$ are like terms.

términos semejante Términos que contienen las mismas variable(s) elevadas a la misma potencia. Ejemplo: $5x$ y $6x$ son términos semejante.

likelihood (Lesson 10-1) The chance of an event occurring.

probabilidad La probabilidad de que ocurra un evento.

linear expression (Lesson 5-2) An algebraic expression in which the variable is raised to the first power, and variables are neither multiplied nor divided.

expresión lineal Expresión algebraica en la cual la variable se eleva a la primera potencia.

linear relationship (Lesson 1-4) A relationship for which the graph is a straight line.

relación lineal Una relación para la cual la gráfica es una línea recta.

M

markdown (Lesson 2-8) An amount by which the regular price of an item is reduced.

rebaja Una cantidad por la cual el precio regular de un artículo se reduce.

markup (Lesson 2-7) The amount the price of an item is increased above the price the store paid for the item.

margen de utilidad Cantidad de aumento en el precio de un artículo por encima del precio que paga la tienda por dicho artículo.

mean (Lesson 11-4) The sum of the data divided by the number of items in the data set.

media La suma de los datos dividida entre el número total de artículos en el conjunto de datos.

mean absolute deviation (MAD) (Lesson 11-4) A measure of variation in a set of numerical data, computed by adding the distances between each data value and the mean, then dividing by the number of data values.

desviación media absoluta Una medida de variación en un conjunto de datos numéricos que se calcula sumando las distancias entre el valor de cada dato y la media, y luego dividiendo entre el número de valores.

measures of center (Lesson 11-4) Numbers that are used to describe the center of a set of data. These measures include the mean, median, and mode.

medidas del centro Números que se usan para describir el centro de un conjunto de datos. Estas medidas incluyen la media, la mediana y la moda.

measures of variation (Lesson 11-4) A measure used to describe the distribution of data.

medidas de variación Medida usada para describir la distribución de los datos.

median (Lesson 11-4) A measure of center in a set of numerical data. The median of a list of values is the value appearing at the center of a sorted version of the list—or the mean of the two central values, if the list contains an even number of values.

mediana Una medida del centro en un conjunto de dados números. La mediana de una lista de valores es el valor que aparace en el centro de una versíon ordenada de la lista, o la media de dos valores centrales si la lista contiene un número par de valores.

monomial (Lesson 5-4) A number, variable, or product of a number and one or more variables.

monomio Número, variable o producto de un número y una o más variables.

Multiplication Property of Equality (Lesson 6-1) If you multiply each side of an equation by the same nonzero number, the two sides remain equal.

propiedad de multiplicación de la igualdad Si multiplicas ambos lados de una ecuación por el mismo número no nulo, lo lados permanecen iguales.

Multiplication Property of Inequality (Lesson 7-3) When you multiply each side of an inequality by a negative number, the inequality symbol must be reversed for the inequality to remain true.

propiedad de desigualdad en la multiplicación Cuando se multiplica cada lado de una desigualdad por un número negativo, el símbolo de desigualdad debe invertirse para que la desigualdad siga siendo verdadera.

Multiplicative Identity Property (Lesson 3-3) The product of any number and one is the number.

propiedad de identidad de la multiplicación El producto de cualquier número y uno es el mismo número.

multiplicative inverse (Lesson 4-5) Two numbers with a product of 1. For example, the multiplicative inverse of $\frac{2}{3}$ is $\frac{3}{2}$.

inverso multiplicativo Dos números cuyo producto es 1. Por ejemplo, el inverso multiplicativo de $\frac{2}{3}$ es $\frac{3}{2}$.

Multiplicative Property of Zero (Lesson 3-3) The product of any number and zero is zero.

propiedad del cero en la multiplicación El producto de cualquier número y cero es cero.

N

negative integer (Lesson 3-1) An integer that is less than zero. Negative integers are written with a − sign.

entero negativo Número menor que cero. Se escriben con el signo −.

net **(Lesson 9-5)** A two-dimensional figure that can be used to build a three-dimensional figure.

red Figura bidimensional que sirve para hacer una figura tridimensional.

nonproportional **(Lesson 1-3)** The relationship between two ratios with a rate or ratio that is not constant.

no proporcional Relación entre dos razones cuya tasa o razón no es constante.

numerical expression **(Lesson 5-1)** A combination of numbers and operations.

expresión numérica Combinación de números y operaciones.

O

obtuse angle **(Lesson 8-1)** Any angle that measures greater than 90° but less than 180°.

ángulo obtuso Cualquier ángulo que mide más de 90° pero menos de 180°.

obtuse triangle **(Lesson 8-3)** A triangle having one obtuse angle.

triángulo obtusángulo Triángulo que tiene un ángulo obtuso.

opposites **(Lesson 3-1)** Two integers are opposites if they are represented on the number line by points that are the same distance from zero, but on opposite sides of zero. The sum of two opposites is zero.

opuestos Dos enteros son opuestos si, en la recta numérica, están representados por puntos que equidistan de cero, pero en direcciones opuestas. La suma de dos opuestos es cero.

order of operations **(Lesson 4-6)** The rules to follow when more than one operation is used in a numerical expression.

1. Evaluate the expressions inside grouping symbols.
2. Evaluate all powers.
3. Multiply and divide in order from left to right.
4. Add and subtract in order from left to right.

orden de las operaciones Reglas a seguir cuando se usa más de una operación en una expresión numérica.

1. Primero, evalúa las expresiones dentro de los símbolos de agrupación.
2. Evalúa todas las potencias.
3. Multiplica y divide en orden de izquierda a derecha.
4. Suma y resta en orden de izquierda a derecha.

outcome **(Lesson 10-1)** Any one of the possible results of an action. For example, 4 is an outcome when a number cube is rolled.

resultado Cualquiera de los resultados posibles de una acción. Por ejemplo, 4 puede ser un resultado al lanzar un cubo numerado.

P

parallelogram **(Lesson 9-3)** A quadrilateral with opposite sides parallel and opposite sides congruent.

paralelogramo Cuadrilátero cuyos lados opuestos son paralelos y congruentes.

percent error **(Lesson 2-4)** A ratio that compares the inaccuracy of an estimate (amount of error) to the actual amount.

porcentaje de error Una razón que compara la inexactitud de una estimación (cantidad del error) con la cantidad real.

percent of change **(Lesson 2-3)** A ratio that compares the change in a quantity to the original amount.

$$\text{percent of change} = \frac{\text{amount of change}}{\text{original amount}} \cdot 100$$

porcentaje de cambio Razón que compara el cambio en una cantidad a la cantidad original.

$$\text{porcentaje de cambio} = \frac{\text{cantidad del cambio}}{\text{cantidad original}} \cdot 100$$

percent of decrease **(Lesson 2-3)** A negative percent of change.

porcentaje de disminución Porcentaje de cambio negativo.

percent of increase **(Lesson 2-3)** A positive percent of change.

porcentaje de aumento Porcentaje de cambio positivo.

pi **(Lesson 9-1)** The ratio of the circumference of a circle to its diameter. The Greek letter π represents this number. The value of pi is 3.1415926.... Approximations for pi are 3.14 and $\frac{22}{7}$.

pi Relación entre la circunferencia de un círculo y su diámetro. La letra griega π representa este número. El valor de pi es 3.1415926.... Las aproximaciones de pi son 3.14 y $\frac{22}{7}$.

plane **(Lesson 8-5)** A two-dimensional flat surface that extends in all directions.

plano Superficie bidimensional que se extiende en todas direcciones.

polygon **(Lesson 9-3)** A simple closed figure formed by three or more straight line segments.

polígono Figura cerrada simple formada por tres o más segmentos de recta.

polyhedron **(Lesson 8-5)** A three-dimensional figure with faces that are polygons.

poliedro Una figura tridimensional con caras que son polígonos.

population **(Lesson 11-1)** The entire group of items or individuals from which the samples under consideration are taken.

población El grupo total de individuos o de artículos del cual se toman las muestras bajo estudio.

positive integer **(Lesson 3-1)** An integer that is greater than zero. They are written with or without a + sign.

entero positivo Entero que es mayor que cero; se escribe con o sin el signo +.

principal **(Lesson 2-9)** The amount of money deposited or borrowed.

capital Cantidad de dinero que se deposita o se toma prestada.

prism **(Lesson 8-5)** A polyhedron with two parallel congruent faces called bases.

prisma Un poliedro con dos caras congruentes paralelas llamadas bases.

probability **(Lesson 10-2)** The chance that some event will happen. It is the ratio of the number of favorable outcomes to the number of possible outcomes.

probabilidad La posibilidad de que suceda un evento. Es la razón del número de resultados favorables al número de resultados posibles.

probability experiment **(Lesson 10-2)** When you perform an event to find the likelihood of an event.

experimento de probabilidad Cuando realiza un evento para encontrar la probabilidad de un evento.

probability model **(Lesson 10-3)** A model used to assign probabilities to outcomes of a chance process by examining the nature of the process.

modelo de probabilidad Un modelo usado para asignar probabilidades a resultados de un proceso aleatorio examinando la naturaleza del proceso.

properties **(Lesson 3-1)** Statements that are true for any number or variable.

propiedades Enunciados que son verdaderos para cualquier número o variable.

proportion **(Lesson 1-5)** An equation stating that two ratios or rates are equivalent.

proporción Ecuación que indica que dos razones o tasas son equivalentes.

proportional (Lesson 1-3) The relationship between two ratios with a constant rate or ratio.

proporcional Relación entre dos razones con una tasa o razón constante.

pyramid (Lesson 8-5) A polyhedron with one base that is a polygon and three or more triangular faces that meet at a common vertex.

pirámide Un poliedro con una base que es un polígono y tres o más caras triangulares que se encuentran en un vértice común.

Q

quadrilateral (Lesson 9-3) A closed figure having four sides and four angles.

cuadrilátero Figura cerrada que tiene cuatro lados y cuatro ángulos.

R

radius (Lesson 9-1) The distance from the center of a circle to any point on the circle.

radio Distancia desde el centro de un círculo hasta cualquiera de sus puntos.

random (Lesson 10-2) Outcomes occur at random if each outcome occurs by chance. For example, rolling a number on a number cube occurs at random.

azar Los resultados ocurren aleatoriamente si cada resultado ocurre por casualidad. Por ejemplo, sacar un número en un cubo numerado ocurre al azar.

rate (Lesson 1-1) A special kind of ratio in which the units are different.

tasa Un tipo especial de relación en el que las unidades son diferentes.

ratio (Lesson 1-1) A comparison between two quantities, in which for every *a* units of one quantity, there are *b* units of another quantity.

razón Una comparación entre dos cantidades, en la que por cada *a* unidades de una cantidad, hay unidades *b* de otra cantidad.

rational numbers (Lesson 4-1) The set of numbers that can be written in the form $\frac{a}{b}$, where *a* and *b* are integers and $b \neq 0$.

Examples: $1 = \frac{1}{1}, \frac{2}{9}, -2.3 = \frac{-23}{10}$

números racionales Conjunto de números que puede escribirse en la forma $\frac{a}{b}$ donde *a* y *b* son números enteros y $b \neq 0$.

Ejemplos: $1 = \frac{1}{1}, \frac{2}{9}, -2.3 = \frac{-23}{10}$

reciprocal (Lesson 4-5) The multiplicative inverse of a number.

recíproco El inverso multiplicativo de un número.

rectangular prism (Lesson 9-4) A prism that has two parallel congruent bases that are rectangles.

prisma rectangular Un prisma con dos bases paralelas congruentes que son rectángulos.

reduction (Lesson 8-4) An image smaller than the original.

reducción Imagen más pequeña que la original.

regular polygon (Lesson 9-3) A polygon that has all sides congruent and all angles congruent.

polígono regular Polígono con todos los lados y todos los ángulos congruentes.

regular pyramid (Lesson 9-5) A pyramid whose base is a regular polygon and in which the segment from the vertex to the center of the base is the altitude.

pirámide regular Pirámide cuya base es un polígono regular y en la cual el segmento desde el vértice hasta el centro de la base es la altura.

relative frequency **(Lesson 10-2)** A ratio that compares the frequency of each category to the total.

frecuencia relativa Razón que compara la frecuencia de cada categoría al total.

relative frequency graph **(Lesson 10-2)** A graph used to organize occurrences compared to a total.

gráfico de frecuencia relativa Gráfico utilizado para organizar las ocurrencias en comparación con un total.

relative frequency table **(Lesson 10-2)** A table used to organize occurrences compared to a total.

tabla de frecuencia relativa Una tabla utilizada para organizar las ocurrencias en comparación con un total.

repeating decimal **(Lesson 4-1)** A decimal in which 1 or more digits repeat.

decimal periódico Un decimal en el que se repiten 1 o más dígitos.

rhombus **(Lesson 9-3)** A parallelogram having four congruent sides.

rombo Paralelogramo que tiene cuatro lados congruentes.

right angle **(Lesson 8-1)** An angle that measures exactly 90°.

ángulo recto Ángulo que mide exactamente 90°.

right triangle **(Lesson 8-3)** A triangle having one right angle.

triángulo rectángulo Triángulo que tiene un ángulo recto.

S

sales tax **(Lesson 2-6)** An additional amount of money charged on items that people buy.

impuesto sobre las ventas Cantidad de dinero adicional que se cobra por los artículos que se compran.

sample **(Lesson 11-1)** A randomly selected group chosen for the purpose of collecting data.

muestra Grupo escogido al azar o aleatoriamente que se usa con el propósito de recoger datos.

sample space **(Lesson 10-3)** The set of all possible outcomes of a probability experiment.

espacio muestral Conjunto de todos los resultados posibles de un experimento probabilístico.

scale **(Lesson 8-4)** The scale that gives the ratio that compares the measurements of a drawing or model to the measurements of the real object.

escala Razón que compara las medidas de un dibujo o modelo a las medidas del objeto real.

scale drawing **(Lesson 8-4)** A drawing that is used to represent objects that are too large or too small to be drawn at actual size.

dibujo a escala Dibujo que se usa para representar objetos que son demasiado grandes o demasiado pequeños como para dibujarlos de tamaño natural.

scale factor **(Lesson 8-4)** A scale written as a ratio without units in simplest form.

factor de escala Escala escrita como una razón sin unidades en forma simplificada.

scale model **(Lesson 8-4)** A model used to represent objects that are too large or too small to be built at actual size.

modelo a escala Réplica de un objeto real, el cual es demasiado grande o demasiado pequeño como para construirlo de tamaño natural.

scalene triangle **(Lesson 8-3)** A triangle having no congruent sides.

triángulo escaleno Triángulo sin lados congruentes.

selling price **(Lesson 2-7)** The amount the customer pays for an item.

precio de venta Cantidad de dinero que paga un consumidor por un artículo.

semicircle **(Lesson 9-2)** Half of a circle. The formula for the area of a semicircle is $A = \frac{1}{2}\pi r^2$.

semicírculo Medio círculo. La fórmula para el área de un semicírculo es $A = \frac{1}{2}\pi r^2$.

simple event **(Lesson 10-2)** One outcome or a collection of outcomes.

eventos simples Un resultado o una colección de resultados.

simple interest **(Lesson 2-9)** The amount paid or earned for the use of money. The formula for simple interest is $I = prt$.

interés simple Cantidad que se paga o que se gana por el uso del dinero. La fórmula para calcular el interés simple es $I = prt$.

simple random sample **(Lesson 11-1)** An unbiased sample where each item or person in the population is as likely to be chosen as any other.

muestra aleatoria simple Muestra de una población que tiene la misma probabilidad de escogerse que cualquier otra.

simplest form **(Lesson 5-1)** An expression is in simplest form when it is replaced by an equivalent expression having no like terms or parentheses.

expresión mínima Expresión en su forma más simple cuando es reemplazada por una expresión equivalente que no tiene términos similares ni paréntesis.

simplify **(Lesson 5-5)** Write an expression in simplest form.

simplificar Escribir una expresión en su forma más simple.

simulation **(Lesson 10-6)** An experiment that is designed to model the action in a given situation.

simulación Un experimento diseñado para modelar la acción en una situación dada.

slant height **(Lesson 9-5)** The height of each lateral face.

altura oblicua Altura de cada cara lateral.

solution **(Lesson 6-1)** A replacement value for the variable in an open sentence. A value for the variable that makes an equation true. Example: The *solution* of $12 = x + 7$ is 5.

solución Valor de reemplazo de la variable en un enunciado abierto. Valor de la variable que hace que una ecuación sea verdadera. Ejemplo: La *solución* de $12 = x + 7$ es 5.

statistics **(Lesson 11-1)** The study of collecting, organizing, and interpreting data.

estadística Estudio que consiste en recopilar, organizar e interpretar datos.

straight angle **(Lesson 8-1)** An angle that measures exactly 180°.

ángulo llano Ángulo que mide exactamente 180°.

stratified random sample **(Lesson 11-1)** A sample in which the population is divided into groups with similar traits that do not overlap. A simple random sample is then selected from each group.

muestra aleatoria estratificada Una muestra en la que la población se divide en grupos con rasgos similares que no se superponen. A continuación, se selecciona una muestra aleatoria simple de cada grupo.

Subtraction Property of Equality **(Lesson 6-1)** If you subtract the same number from each side of an equation, the two sides remain equal.

propiedad de sustracción de la igualdad Si restas el mismo número de ambos lados de una ecuación, los dos lados permanecen iguales.

Subtraction Property of Inequality (Lesson 7-1) If you subtract the same number from each side of an inequality, the inequality remains true.

propiedad de desigualdad en la resta Si se resta el mismo número a cada lado de una desigualdad, la desigualdad sigue siendo verdadera.

supplementary angles (Lesson 8-2) Two angles are supplementary if the sum of their measures is 180°.

ángulos suplementarios Dos ángulos son suplementarios si la suma de sus medidas es 180°.

surface area (Lesson 9-5) The sum of the areas of all the surfaces (faces) of a three-dimensional figure.

área de superficie La suma de las áreas de todas las superficies (caras) de una figura tridimensional.

survey (Lesson 11-1) A question or set of questions designed to collect data about a specific group of people, or population.

encuesta Pregunta o conjunto de preguntas diseñadas para recoger datos sobre un grupo específico de personas o población.

symmetric distribution (Lesson 11-4) A distribution in which the shape of the graph on each side of the center is similar.

distribución simétrica Distribución en la que la forma de la gráfica en cada lado del centro es similar.

systematic random sample (Lesson 11-1) A sample where the items or people are selected according to a specific time or item interval.

muestra aleatoria sistemática Muestra en que los elementos o personas se eligen según un intervalo de tiempo o elemento específico.

T

term (Lesson 5-1) A number, a variable, or a product or quotient of numbers and variables.

término Número, variable, producto o cociente de números y de variables.

terminating decimal (Lesson 4-1) A decimal with a repeating digit of 0.

decimal finito Un decimal que tiene un dígito que se repite que es 0.

theoretical probability (Lesson 10-3) The ratio of the number of ways an event can occur to the number of possible outcomes in the sample space. It is based on what *should* happen when conducting a probability experiment.

probabilidad teórica Razón del número de maneras en que puede ocurrir un evento al número de resultados posibles en el espacio muestral. Se basa en lo que *debería* pasar cuando se conduce un experiment probabilístico.

theoretical probability of a compound event (Lesson 10-5) The ratio of the number of ways an event can occur to the number of possible outcomes in the sample space. It is based on what *should* happen when conducting a probability experiment.

probabilidad teórica de un evento compuesto Razón del número de maneras en que puede ocurrir un evento al número de resultados posibles en el espacio muestral. Se basa en lo que *debería* pasar cuando se conduce un experimento probabilístico.

three-dimensional figure (Lesson 8-5) A figure with length, width, and height.

figura tridimensional Figura que tiene largo, ancho y alto.

tip (Lesson 2-7) Also known as a gratuity, it is a small amount of money in return for a service.

propina También conocida como gratificación; es una cantidad pequeña de dinero en recompensa por un servicio.

trapezoid (Lesson 9-3) A quadrilateral with one pair of parallel sides.

trapecio Cuadrilátero con un único par de lados paralelos.

tree diagram **(Lesson 10-5)** A diagram used to show the sample space.

diagrama de árbol Diagrama que se usa para mostrar el espacio muestral.

triangle **(Lesson 8-3)** A figure with three sides and three angles.

triángulo Figura con tres lados y tres ángulos.

triangular prism **(Lesson 9-4)** A prism that has two parallel congruent bases that are triangles.

prisma triangular Un prisma que tiene dos bases congruentes paralelas que triángulos.

two-step equation **(Lesson 6-2)** An equation having two different operations.

ecuación de dos pasos Ecuación que contiene dos operaciones distintas.

two-step inequality **(Lesson 7-6)** An inequality that contains two operations.

desigualdad de dos pasos Desigualdad que contiene dos operaciones.

U

unbiased sample **(Lesson 11-1)** A sample representative of the entire population.

muestra no sesgada Muestra que se selecciona de modo que se representativa de la población entera.

uniform probability model **(Lesson 10-3)** A probability model which assigns equal probability to all outcomes.

modelo de probabilidad uniforme Un modelo de probabilidad que asigna igual probabilidad a todos los resultados.

unit rate **(Lesson 1-1)** A rate in which the first quantity is compared to 1 unit of the second quantity.

tasa unitaria Una tasa en la que la primera candidad se compara con 1 unidad de la segunda candidad.

unit ratio **(Lesson 1-2)** A ratio in which the first quantity is compared to every 1 unit of the second quantity.

razón unitaria Una relación en la que la primera cantidad se compara con cada 1 unidad de la segunda cantidad.

V

valid inference **(Lesson 11-1)** A prediction, made about a population, based on an unbiased sample that is representative of the population.

inferencia válida Una predicción, hecha sobre una población, basada en una muestra imparcial que es representativa de la población.

valid sampling method **(Lesson 11-1)** A sampling method that is: representative of the population selected at random, where each member has an equal chance of being selected, and large enough to provide accurate data.

método de muestreo válido Un método de muestreo que es: representativo de la población seleccionada al azar, donde cada miembro tiene la misma oportunidad de ser seleccionado y suficientemente grande para proporcionar datos precisos.

variability **(Lesson 11-3)** A measure that describes the amount of diversity in values within a sample or samples.

variabilidad Medida que describe la cantidad de diversidad en valores dentro de una muestra o muestras.

variable **(Lesson 5-1)** A symbol, usually a letter, used to represent a number in mathematical expressions or sentences.

variable Símbolo, por lo general una letra, que se usa para representar un número en expresiones o enunciados matemáticos.

vertex (Lesson 8-1) A vertex of an angle is the common endpoint of the rays forming the angle.

vértice El vértice de un ángulo es el extremo común de los rayos que lo forman.

vertex (Lesson 8-5) The point where three or more faces of a polyhedron intersect.

vértice El punto donde tres o más caras de un poliedro se cruzan.

vertical angles (Lesson 8-1) Opposite angles formed by the intersection of two lines. Vertical angles are congruent.

ángulos opuestos por el vértice Ángulos opuestos formados por la intersección de dos rectas. Los ángulos opuestos por el vértice son congruentes.

vertices (Lesson 8-5) Plural of vertex.

vértices Plural de verticé.

visual overlap (Lesson 11-4) A visual demonstration that compares the centers of two distributions with their variation, or spread.

superposición visual Una demostración visual que compara los centros de dos distribuciones con su variación, o magnitud.

volume (Lesson 9-4) The number of cubic units needed to fill the space occupied by a solid.

volumen Número de unidades cúbicas que se requieren para llenar el espacio que ocupa un sólido.

voluntary response sample (Lesson 11-1) A sample which involves only those who want to participate in the sampling.

muestra de respuesta voluntaria Muestra que involucra sólo aquellos que quieren participar en el muestreo.

W

wholesale cost (Lesson 2-7) The amount the store pays for an item.

coste al por mayor La cantidad que la tienda paga por un artículo.

Z

zero angle (Lesson 8-1) An angle that measures exactly 0 degrees.

ángulo cero Un ángulo que mide exactamente 0 grados.

zero pair (Lesson 3-1) The result when one positive counter is paired with one negative counter. The value of a zero pair is 0.

par nulo Resultado de hacer coordinar una ficha positiva con una negativa. El valor de un par nulo es 0.

Index

A

B

C

D

E

Lesson 1-1 Unit Rates Involving Ratios of Fractions, Practice Pages 11-12

1. 64 miles in one hour **3.** 124.8 miles in one hour **5.** $\frac{5}{9}$ feet in one month **7.** $\frac{3}{5}$ mile in one minute **9.** penny; about 56.4 kilometers per hour faster **11.** Sample answer: The student who runs $\frac{3}{4}$ mile in 6 minutes will run 1 mile in 8 minutes if the rate is constant. The student who runs $\frac{1}{2}$ mile in 5 minutes will run 1 mile in 10 minutes if the rate is constant. **13.** Sample answer: She set up her rate as $\frac{\frac{3}{4}\text{ hour}}{9\text{ greeting cards}}$. Her unit rate of $\frac{1}{12}$ is hour per greeting card, not card per hour.

Lesson 1-2 Understand Proportional Relationships, Practice Pages 19-20

1. yes; Both have a ratio of 3 : 1. **3.** yes; Both have a ratio of 7 : 200. **5.** no; One ratio is 4 cm to 1 year and the other is 3 cm to 1 year. **7.** 5 ft^3 minerals, 5 ft^3 peat moss, 10 ft^3 compost; 12 ft^3 minerals, 12 ft^3 peat moss, 24 ft^3 compost; 20 ft^3 minerals, 20 ft^3 peat moss, 40 ft^3 compost; 100 ft^3 minerals, 100 ft^3 peat moss, 200 ft^3 compost **9.** $\frac{1}{2}$ cup **11.** Sample answer: In the first cleaning solution, the ratio of vinegar to water is 1 : 2. The second solution, however, has a ratio of 2 : 3. The ratios are not equivalent. **13.** Sample answer: The unit rate for each ratio of a situation is the quantity compared to 1 unit of another quantity. If a relationship is proportional, then each ratio would have the same unit rate.

Lesson 1-3 Tables of Proportional Relationships, Practice Pages 29-30

1.

Lunches Bought	1	2	3	4
Total Cost ($)	2.50	5.00	7.50	10.00

yes; The ratios between the quantities are all equal and have a unit rate of $2.50 per lunch.

3.

Hours	1	2	3	4
Cost ($)	55	75	95	115

no; The ratios between the quantities are not equal.

5. 0.50 **7.** 3.1 **9.** yes; $9.00 **11.** yes; Sample answer: There is a proportional relationship between the number of fluid ounces and the number of cups. So, the number of cups would increase by the same factor as the number of fluid ounces. **13.** Sample answer: The number of laps increases at the same rate and the time increases at the same rate, but the ratios of laps to time are not equal.

Lesson 1-4 Graphs of Proportional Relationships, Practice Pages 39-40

1.

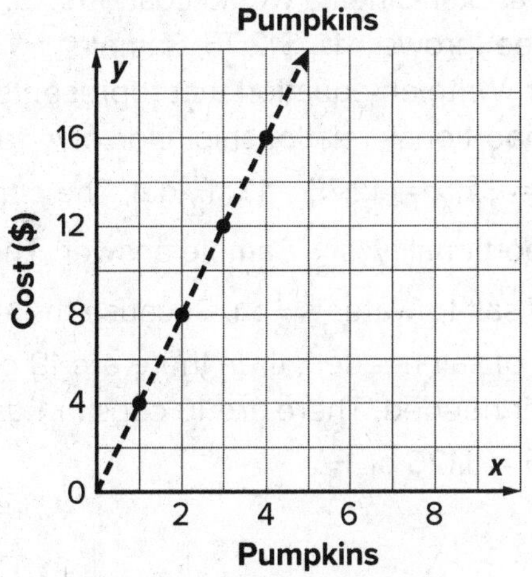

The cost is proportional to the number of pumpkins bought because the graph is a straight line through the origin.

3. 15

5.

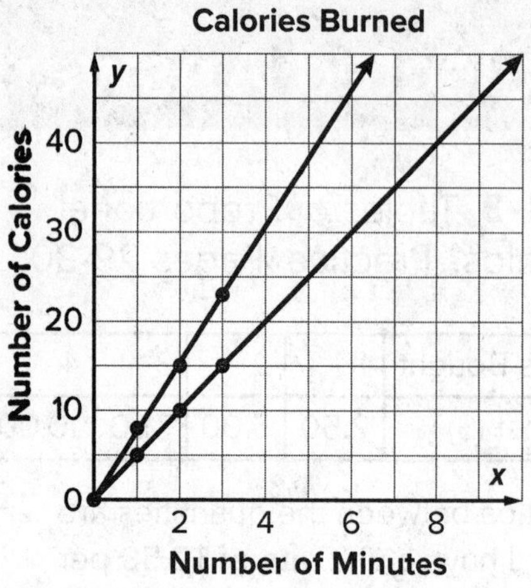

The point (1, 5) represents the unit rate.

7. Sample answer: A line can have constant unit rate and not be proportional because it does not pass through the origin. For example, renting a boat where there is an initial fee and then a cost per hour. This graph would have a constant rate but not pass through the origin.
9. Sample answer: The graph shows the number of feet per yard. So, for every yard there are 3 feet.

Lesson 1-5 Equations of Proportional Relationships, Practice Pages 47-48

1. 4.75; Liv earns $4.75 for each bracelet she sells. **3.** $y = 1.5x$ **5.** $y = 1\frac{3}{4}x$; $10\frac{1}{2}$ c
7. The equation $y = 7.75x$ models the cost for tickets at Star Cinema. The total cost of 9 tickets at Star Cinema would cost $69.75.
9. 91 free throws **11.** $12.75; Sample answer: Write an equation that represents the proportional relationship and solve for x. $38.25 = 3x$; $x = 12.75$ **13.** $1\frac{1}{8}$ gal; The constant of proportionality is $\frac{1}{3}$. Sample answer: The ratio of salt to water is $\frac{1}{3}$: 1. Because there are 6 cups of salt needed, then there are 18 cups of water needed. There are 16 cups in 1 gallon. $18 \div 16 = 1.125$ or $1\frac{1}{8}$.

Lesson 1-6 Solve Problems Involving Proportional Relationships, Practice Pages 55-56

1. 177 girls **3.** 75 pieces **5.** 19.2 min
7. 315 adults **9.** 0.52 gal; Sample answer: The area of the fence is 26(7) or 182 square feet. Using the equation $y = 350x$ to represent the proportional relationship, she will need $182 \div 350$ or 0.52 gal. **11.** Sample answer: Enrique goes through a 12-ounce bottle of shampoo in 16 weeks. How long would you expect an 18-ounce bottle of the same brand to last him? 24 weeks

Module 1 Review Pages 59-60

1. D **3.** 8 cups flour, 2 cups salt, 4 cups water; 6 cups flour, $1\frac{1}{2}$ cups salt, 3 cups water
5. no; The ratios $\frac{2}{1}$, $\frac{5}{4}$, $\frac{7}{6}$, and $\frac{10}{8}$ are not equivalent ratios. **7.** The relationship is proportional. The point (9, 18) satisfies this relationship. The constant of proportionality is 2. **9A.** $y = 22.5x$ **9B.** $292.50

Lesson 2-1 Percent of Change, Practice Pages 71-72

1. 25%; increase **3.** 37.5%; decrease
5. 48.5% **7.** 10.7% **9.** 56.6% **11.** area; 600%
13. Sample answer: Last month the amount of snowfall was 0.5 inch. This month it was 0.75 inch. What is the percent of increase in snowfall?; 50% **15.** true; Sample answer: A decrease indicates a lesser amount. So, the original amount must be more than the decreased amount.

Lesson 2-2 Tax, Practice Pages 81-82

1. \$19.26 **3.** \$52.45 **5.** \$79.82 **7.** \$359.13 **9.** D. \$132.74 **11.** \$34.54 **13.** Sample answer: $1.06 \cdot a$; $a + 0.06a$; Both expressions multiply the cost by 100% and by the sales tax rate to give the total cost. **15.** more; Sample answer: 10% of \$160 is \$16 so 5% of \$160 is \$8. The tax rate is greater than 5% so the tax will be more than \$8.

Lesson 2-3 Tips and Markups, Practice Pages 89-90

1. \$22.00 **3.** \$194.25 **5.** \$47.40 **7.** \$30 **9.** 55% **11.** \$342.37 **13.** yes; Sample answer: The total would be \$77 which is less than \$80. **15.** Sample answer: $1.18x$; $x + 0.18x$; Both expressions multiply the cost by 100% and by the tip to give the total cost.

Lesson 2-4 Discounts, Practice Pages 97-98

1. \$126.00 **3.** \$276.25 **5.** \$24.70 **7.** \$375.15 **9.** \$47.89 **11.** Skyebouncer; \$3.76; **13.** $11\frac{1}{9}$%; Sample answer: Suppose the cost of the item is \$50. So, 10% of \$50 is \$5 and \$50 − \$5 = \$45. Then set up a proportion to find what percent of \$45 is \$5. Solve for r: $\frac{5}{45} = \frac{r}{100}$. $r = 11\frac{1}{9}$% **15.** \$34.30; about 14%

Lesson 2-5 Interest, Practice Pages 105-106

1. \$31.80 **3.** \$157.50 **5.** \$112.50 **7.** \$288.75 **9.** \$157.50 **11.** \$247.50 **13.** 3.75%; Use the simple interest formula and solve for r. $32{,}813 = 2{,}500 \cdot 3.5 \cdot r$; $r = 3.75$ **15.** Sample answer: If the rate is doubled, then the interest is doubled to \$80. If the time is doubled, then the interest is also doubled to \$80.

Lesson 2-6 Commission and Fees, Practice Pages 113-114

1. \$391 **3.** \$27,273 **5.** \$20,000 **7.** \$12 **9.** Job 2; \$108.75 **11.** Sample answer: The student wrote the percent as 65 out of 100. It should be 6.5 out of 100. The commission should be \$34.13. **13.** true; Sample answer: you could earn 0.5% of a sale of \$5,000 worth of jewelry. 0.5% of \$5,000 is \$25.

Lesson 2-7 Percent Error, Practice Pages 119-120

1. 30% **3.** 36.9% **5.** 1.01% **7.** 34.1% **9.** 4% **11.** The student forgot to write the amount of error as a percent. 0.12 = 12% **13.** The denominator would be 0 in the calculation. A denominator of 0 is undefined so the percent of error would be undefined.

Module 2 Review Pages 123-124

1. 48% **3.** B

5.

	yes	no
subtotal: \$129.50 tip: \$23.27		x
subtotal: \$98.40 tip: \$17.71	x	
subtotal: \$142.17 tip: \$22.75		x

7. 47.19 **9.** There are actually 118 elements. The amount of error in Clark's estimate is 10. To the nearest tenth percent, the percent error in Clark's estimate is 8.5%. **11.** \$64,815; I solved the proportion $\frac{3{,}500}{x} = \frac{5.4}{100}$.

Selected Answers

Lesson 3-1 Add Integers, Practice Pages 137-138

1. −11 **3.** −26 **5.** −17 **7.** −40 **9.** −20 **11.** −1 **13.** 76 m higher **15.** Tasha; $1 **17a.** −1 **17b.** 3 **17c.** 0 **19.** Sample answer: 2 and −2 are additive inverses and the sum of any number and its additive inverse is zero. The integer 3 is positive, so the sum will be positive.

Lesson 3-2 Subtract Integers, Practice Pages 147-148

1. 11 **3.** 76 **5.** −10 **7.** −6 **9.** 22 **11.** −37 **13.** 19 units **15.** 440 feet **17.** Utah; Nevada **19.** The student incorrectly wrote 4 − 2 instead of 4 + 2. The correct solution is 6. **21.** sometimes; Sample answer: For example, −10 − (−40) = 30 and −28 − (−13) = −15.

Lesson 3-3 Multiply Integers, Practice Pages 157-158

1. −28 **3.** −108 **5.** 100 **7.** −140 **9.** −192 **11.** 70 **13.** 756 **15.** −63 points **17.** yes; At the end of January, he had $1,300 −$1,250 or $50 left in his account. $50 • 12 = $600 and $600 > $500. **19a.** Multiplicative Identity Property **19b.** Commutative Property of Multiplication **21.** −8 and 8

Lesson 3-4 Divide Integers, Practice Pages 165-166

1. −11 **3.** −25 **5.** 3 **7.** −16 **9.** −4 **11.** 3 **13.** −60 **15.** 16 weeks **17.** no; Sample answer: The Associative Property is not true for the division of integers because the way the integers are grouped affects the solution. [12 ÷ (−6)] ÷ 2 = −1; 12 ÷ [(−6 ÷ 2)] = −4 **19.** Sample answer: Lucy borrowed $50 from her mother over 5 days in equal amounts. What was the change in the amount she borrowed from her mom each day? −$10

Lesson 3-5 Apply Integer Operations, Practice Pages 169-170

1. −39 **3.** 22 **5.** −105 **7.** −6 **9.** −7 **11.** 10 **13.** −$200 **15.** Chicago; Oklahoma City; The difference between the extremes for Chicago is 40°C − (−33°C) or 77°C. The difference between the extremes for Nashville is 42°C − (−27°C) or 69°C. The difference between the extremes for Oklahoma City is 43°C − (−22°C) or 65°C. 77 > 69 > 65 **17.** The student subtracted 9 instead of adding its additive inverse. The correct answer should be −63.
19. no; According to the order of operations, multiplication should be performed from left to right before subtraction.

Module 3 Review Pages 173-174

1. +3 strokes to par **3A.** Commutative Property, Sample answer: One of the methods that can be used to add three or more integers is to group like signs together, then add. **3B.** 35 **5.** C **7.** 200

9.

	positive	negative
−14 ÷ (−2)	x	
$\frac{64}{-4}$		x
1,256 ÷ 8	x	

11. C

Lesson 4-1 Rational Numbers, Practice Pages 183-184

1. 0.625; terminating **3.** $0.\overline{2}$; non-terminating **5.** −0.8; terminating **7.** $-0.\overline{409}$; non-terminating **9.** $-0.\overline{03}$; non-terminating **11.** 0.21875; terminating **13A.** 0.3; terminating **13B.** $0.0\overline{3}$; non-terminating **15.** Cho, Kevin, Sydney **17.** no; Sample answer: $0.\overline{5} = \frac{5}{9}$ **19.** $0.\overline{2}$, $0.\overline{50}$, and $0.\overline{98}$; Sample answer: When the denominator of the fraction is 9 or 99, the numerator of the fraction is the repeating part of the decimal.

Lesson 4-2 Add Rational Numbers, Practice Pages 195-196

1. $\frac{1}{2}$ **3.** $-\frac{9}{10}$ **5.** \$37.20; Sample answer: Quinn gave \$37.20 to his brother for his mother's birthday gift. **7.** $2\frac{2}{3}$ or $2.\overline{6}$ **9.** $1\frac{17}{24}$ or $1.708\overline{3}$ **11.** −3.45 or $-3\frac{9}{20}$ **13.** −0.05 or $-\frac{1}{20}$ **15.** 20.8 in. or $20\frac{4}{5}$ in. **17.** 55 pounds **19.** Sample answer: $3\frac{1}{2} + 1\frac{1}{16}$; $4\frac{9}{16}$ **21.** Sample answer: The student found a common denominator but not the least common denominator. The least common denominator of 9, 3, and 6 is 18.

Lesson 4-3 Subtract Rational Numbers, Practice Pages 201-202

1. 1.45 or $1\frac{9}{20}$ **3.** 8.27 or $8\frac{27}{100}$ **5.** $11\frac{1}{6}$ or $11.1\overline{6}$ **7.** $-3\frac{1}{24}$ or $-3.041\overline{6}$ **9.** $-5\frac{1}{10}$ or −5.1 **11.** $-\frac{1}{6}$ or $-0.1\overline{6}$ **13.** 1.55 or $1\frac{11}{20}$ **15.** 2 or $\frac{2}{1}$ **17.** −5.25 ft or $-5\frac{1}{4}$ ft **19.** false; Sample answer: $2\frac{1}{2} - \left(-2\frac{1}{4}\right) = 4\frac{3}{4}$ **21.** Sample answer: You will get the correct answer but you will need to simplify at the end.

Lesson 4-4 Multiply Rational Numbers, Practice Pages 211-212

1. $\frac{2}{5}$ **3.** $\frac{1}{5}$ **5.** $-7\frac{9}{16}$ **7.** $-\frac{2}{5}$ or −0.4 **9.** $2\frac{3}{20}$ or 2.15 **11.** −1 or $-\frac{1}{1}$ **13.** 7.98 or $7\frac{49}{50}$ **15.** −\$3.13 **17.** Sample answer: $\frac{2}{3}\left(\frac{1}{3}\right) = \frac{2}{9}$ and $\frac{2}{9}$ is $0.\overline{2}$, which is less than 0.25 or $\frac{1}{4}$. **19a.** $\frac{1}{6}$ **19b.** $\frac{1}{8}$

Lesson 4-5 Divide Rational Numbers, Practice Pages 219-220

1. −4 or $-\frac{4}{1}$ **3.** $-1\frac{1}{2}$ **5.** $\frac{4}{5}$ **7.** $-2\frac{11}{17}$ **9.** $-2\frac{2}{7}$ **11.** −6 or $-\frac{6}{1}$ **13.** −8 or $-\frac{8}{1}$ **15.** $3\frac{4}{5}$ or 3.8 **17.** \$228.15 **19.** $20 \div \frac{3}{4}$; Multiplying 20 by a number less than 1 will result in a number less than 20. Dividing 20 by a number less than 1 will result in a number greater than 20. **21.** false; Sample answer: $-\frac{4}{5} \div \left(-\frac{8}{9}\right) = \frac{9}{10}$ and $-\frac{8}{9} \div \left(-\frac{4}{5}\right) = 1\frac{1}{9}$. $\frac{9}{10} \neq 1\frac{1}{9}$.

Lesson 4-6 Apply Rational Numbers Operations, Practice Pages 225-226

1. −2.55 or $-2\frac{11}{20}$ **3.** −1.825 or $-1\frac{33}{40}$ **5.** 0.5 or $\frac{1}{2}$ **7.** −1.85 or $-1\frac{17}{20}$ **9.** 0.26 or $\frac{13}{50}$ **11.** 0.28 or $\frac{7}{25}$ **13.** 33.6 **15.** $\frac{3}{4}$ cup oats; $1\frac{7}{8}$ cups of blueberries **17.** $\left(\frac{1}{2} \times 120 \times 0.25\right) + \left(\frac{1}{5} \times 120 \times 0.75\right) + \left(\frac{3}{10} \times 120 \times 1.50\right)$; \$87
19. Sample answer: A homeowner is enclosing his rectangular property with fencing. The rectangular property is $25\frac{1}{3}$ yards wide and $30\frac{2}{3}$ yards long. Fencing is sold in 8-foot sections and costs \$45.50 per section. How much will it cost to fence in the rectangular property? \$1,911

Module 4 Review Pages 229-230

1. 0.85 **3A.** −\$40.75

3B.

	yes	no
losing \$40.75	X	
donating \$40.75	X	
finding \$40.75		X
earning \$40.75		X
spending \$40.75	X	
receiving a gift of \$40.75		X

5. 37.83 **7.** 8.375 or $8\frac{3}{8}$ in^2; If I find the areas of triangles having the same height with bases 5 and 6, their areas are 7.5 in^2 and 9 in^2, respectively. My answer is about halfway between those areas, so I know my answer is reasonable. **9.** D **11.** \$47.25; Sample answer: I divided 31.5 by the thickness of the DVD case, 0.6 to get 52.5. So, each shelf can hold at most 52 DVDs. Then I divided 132 by 52 to get approximately 2.5. Therefore, I know that I need 3 shelves. Then I multiplied 3 by the cost per shelf, \$15.75.